Fire-Gazing

When Venus Transits the Sun
2004 and 2012

Gail B. Dimitroff, PhD

Order this book online at www.trafford.com
or email orders@trafford.com

Most Trafford titles are also available at major online book retailers.

Printed in the United States of America.

ISBN: 978-1-4669-1677-7 (sc)
ISBN: 978-1-4669-1679-1 (hc)
ISBN: 978-1-4669-1678-4 (e)

Library of Congress Control Number: 2012903187

Trafford rev. 03/05/2012

 www.trafford.com

North America & international
toll-free: 1 888 232 4444 (USA & Canada)
phone: 250 383 6864 ♦ fax: 812 355 4082

Table Of Contents

List of Illustrations

List Of Tables

I dedicate this book to Robbie Adkins. Without her encouragement and support it would never have been written.

Foreward I

"To express freely, released by beauty into the eternal light where truth is perceived—this is our destiny. Love is the way."

Early in *Fire-Gazing; When Venus Transits the Sun*, Gail Dimitroff shares these potent words about Humanity's journey. And throughout this entire excellent book, she reminds us that the transits of Venus are not simply fascinating astronomical phenomena but opportunities to lead with love and to bring a powerful peace to our planet.

I have known Gail Dimitroff for many years. We are students of the books of Alice Bailey and Djwal Khul and have attended many conferences and classes together. Her insights throughout our study always struck a deep chord of truth. Her words penetrate and inspire. She is a visionary and an activist. She acts, writes and speaks for Humanity's growth and for peace on our planet.

No wonder she took it upon herself to explore the Transits of Venus. Here she delves into the actual astronomical events, their astrological implications and the social, economical, artistic and political awakenings that occur as a result of these transits. She looks to the past few hundred years to illumine the beauty Venus has already bestowed in the early transits to the Sun and she looks forward to the grace, peace and soul revelation possible if Humanity prepares for and understands the upcoming opportunity present in the transit of June 2012.

In Gail's exploration of the previous transits of Venus to the Sun, she discusses and explores the astrological charts of outstanding contributors to the world of thought, art and historical trends. The Venusian transits call in potent souls who have deeply affected the flowering of our civilization.

Gail writes, "A critical mass of inspired souls can have a magnificent effect on the whole of humanity." And later, "An infusion of joy, beauty and peace within the collective consciousness is bound to affect the realization that there are no individual problems. This kind of planetary awareness can only come about as we accept that we are all united in One Human Family."

As Gail points out, the Venus transit is relatively short in duration, but "deserves to be examined because it suggests a social cycle of change and transformation . . . a transformation that requires illumination to take us out of our present destructive and inefficient modes into one that will renew the planet and all her inhabitants."

Fire Gazing is a call to end the age of separateness and begin to work together with Love and Light and Power. The Venus Transits particularly work to awaken Humanity's Heart and remind us what is *truly* valuable.

Fire Gazing asks us to awaken to the beauty in our own lives and reveal that beauty as an act of generosity and an affirmation of life. As we choose to share our beauty and appreciate beauty within others, "a world-shaking empathy" as Gail writes, will be released and a New Age will dawn.

Heidi Rose Robbins
Los Angeles 2012

Foreward II

Scientific understanding, soul insight and soul wisdom are the guiding lights comprehensively informing Gail Dimitroff's most recent book, Firegazing, the Transits of Venus 20042012. Gail has synthesized into a meaningful whole the Venus transits from the time of Galileo to the present global context through the lenses of esoteric astrology, science (esoteric and concrete), mythology, history (esoteric and secular), and not least, spiritually. Gail draws out the more relevant aspects of the Venus transits for our learning and application. This book is also a guide through a portal into the future as she "sees into" what are probable opportunities arising from the most recent Venus transit period (2004 and 2008) that Humanity faces for long term choices that are in line with the underlying Plan of God. In the end, the reader will experience Venus brought to life, not as a mass of hot fumes unsuitable for physical living, nor as an impersonal data field of astrological facts and aspects, but as a luminous, blessed Being of universal Beauty, a Presence so refined She blesses all the lives within Her influence, emanating the living qualities of the Sacred Soul She Is. Magnified by the beauteous orb's proximity, Venus inspires the gazer to reflect upon one's own interior Self luminosityradiating love, gentle beauty, shining intelligent clarity, and magnetic Soul power. Enjoy the read!

Halina Bak-Hughes, MS.E

Preface

The title of this book should be explanatory as to its intent. It is an attempt to explain the conditions necessary for a Venus transit to be fulfilled. Secondarily, it is an investigation into the deeper symbolic meanings of the transits of 2004 and 2012. The tools of both exoteric and esoteric astrology will be employed as well as an assessment of past occurrences—including a peek into the history of art, music, and human events—that are coincidental to the previous transits that have been cataloged. This is not an extensive catalog. It focuses primarily on Western Civilization, a circumstance pointing to my own limitation. I am not, unfortunately, a student of the East and I am sure that there are many fascinating and important examples from Asia, India as well as Africa that would enhance this compilation. Readers are encouraged to participate in this study and to forward such information to the e-mail address at the end of the introduction.

It is not necessary to be an astrologer to comprehend or appreciate this analysis. Anyone who is interested in the history of human consciousness, its growth, and development, will find some usefulness in this treatise.

I have tried to explain the importance of the planet Venus in our lives as well as the additional energy opportunity that comes our way with the transits of Venus 2004 and 2012. This explanation has been couched in terms of both exoteric and esoteric astrology in order to deepen the analysis and to present concepts that are critical to humanity at this time. I have described some cyclic laws that apply from both an astronomical and astrological perspective in an attempt to bring about some kind of synthesis. Additionally, I have used the Sabian symbols when and where I had confidence in the accuracy of the degree in question. Where I had uncertainty, I left it out.

My formal education is in Leadership/Human Behavior. I am interested in those individuals who respond in a concrete, meaningful way to the available energies in the direction of true leadership—

leadership that is not selfserving but strives for the evolution of humanity and the greatest good for the greatest number.

The cycles of the Venus transits have been listed in Chapter One. Each cycle or set of transits presents certain themes. They will then be worked out over the next 105 years when a new set of circumstances and themes will present. They are like book ends in time—one in 2004 and the next in 2012. The chess board is set, the pieces in place and the game will play itself out.

These cycles refer more to social change and do not so much focus specifically on individuals to any great extent. Nevertheless, I have included the astrological charts of many individuals from the world of art, music, and history to see how these individuals, who were trend setters, exemplify the era or shed light on some particular astrological aspect. For example, the work of the Impressionist painters came into ascendancy during the two 19th century transits. Their influence upon color and light changed the world of art forever bringing to public awareness the fragility of the moment and the ever changing nature of reality. This approach diverged significantly from the paintings of the past, and put in place a new paradigm.

The various horoscopes are cited so that we may see how individuals fit into the plan and how they respond and reflect the incoming energy. I am interested in leadership that furthers the purpose and the plan inherent in the logos—that logical yet mysterious ordering principle that has fascinated humanity since the beginning of time.

Experience has taught me that there are a myriad of clues in the heavens that can be used to shed light upon our way. Fire-gazing is not only a fascinating pastime but also a guide to human growth and evolution. A word of warning—anyone who takes time to engage in fire-gazing, looking at the sun, must be sure to protect the eyes. The soul will take care of itself.

Gail R. Dimitroff, PhD
January 23, 2012
gaildimitroff@cox.net

Warning: To view the transits of Venus it is imperative to protect the eyes. Even a short glimpse with the naked eye could cause serious damage. A filter should be used with a telescope. In any event never look at the sun directly through a telescope, not even the view finder. Plan ahead to avoid injury. Sun glasses do not provide adequate protection. A number 14 welder's glass available at hardware stores is considered to be a safe filter.

Acknowledgments

I want to thank Michael Robbins for the long hours that he spent teaching people like me esoteric astrology. We would go on for a solid week from early morning to very late at night. During the day we learned special musical compositions that enhanced the teaching—they were used to reinforce the knowledge. On the last night of each session, we proudly performed the music we had been practicing all week in class. I think that sometimes Michael must have winced at our lack of talent for he had been an opera conductor, but he never let on. These classes lasted for five years—a truly unforgettable experience. On the last night we always held a performance and sang our heats out. Michael never asked for anything but attention, seriousness and commitment to the learning. He was then and still is an invaluable mentor, friend and guide. Michael stimulated my interest in esoteric astrology and it endures today as a tool for better grasping the challenges of the times. More than that, and on a much larger scale, Michael models loving understanding, good will, compassion and creative expression.

Glenda Christian must be mentioned. She is a friend, a mentor and solid critic. Her comments on this book were invaluable. She offered fresh insights and caught some errors that I had missed. Her support in this endeavor as well as her continuing support on the path are deeply appreciated and duly noted. Whenever I have a question, Glenda is the first person I think of. She always has a good and practical answer as well as much needed psychological support.

I also want to acknowledge Malvin Artley. It was he who first stimulated my interest in the planet Venus and gave me the term "Firegazing"—a word I think he invented. I liked the word so much I could not let go of it and asked if I could use it. He said, "yes." Once I started to study Venus and the transits, I delved deeper and deeper. Each month Malvin sends out a report on the Full Moon of that period. His insights are fresh, deep and meaningful. I deeply appreciate and acknowledge his constancy of purpose and skill in delivery. Moreover, Malvin is the only person that understands and explains Chinese astrology in a way that makes any sense to me. It is intriguing and although often baffling—still I want to know more.

Halina Bak Hughes has long been a supporter of my work and effort. Over many years, we have spent many long hours in class together pondering the work and supporting group effort. Her dedicated effort at the University of the Seven Rays and its Conference is a great service for all interested in the Ancient Wisdom. I am deeply grateful to Halina for understanding my deep interest in culture, art, politics and astrology—as a unified field of study. She wholeheartedly accepts and fortifies my belief in women and their outstanding efforts for peace and justice.

Heidi Rose Heffner is not only a wonderful astrologer but also a true white magician. Her poetry evokes new and powerful images, her workshops break barriers and her astrology is a magnificent blend of the exoteric and the esoteric. I will always remember her "Hello Love" project with joy, amazement and satisfaction. She loves women, promotes women and understands their power and value to the world of today. Being in her presence in like feeling the heat of a radiant, glowing heart.

Lastly, I want to thank my husband, Tony Dimitroff. He not only reads and proofs my manuscript, but also tends to all computer issues no matter how big or small. He supports my work totally and is a grand and steady companion on the path that we have chosen to walk together. He encourages my work, assists with classes and has never missed a Full Moon session in over thirty years. His support makes the work possible. To him and to all mentioned, I am deeply grateful.

Gail Dimitroff

"When the shutters of selfhood are being fastened everywhere, when fires are extinguished in the darkness, is it not the time to reflect about the Beautiful?"
-Fiery World #177

Chapter One

FIRE-GAZING

When the June flowers are blooming in 2004
Harkness

What happened when, on June 8, 2004, Venus orbited across the face of the sun? There were no transits of Venus during the twentieth century, so this rare event was eagerly anticipated by astronomers, astrologers, and amateur stargazers alike. Some people had been planning for years to catch sight of this event of rare beauty, knowing that the next one will not occur until 2012 and then again not for over one hundred years.

To be exact, the transits occur at intervals of 105.5 years, 8 years, 121.5 years, and 8 years, so it is no wonder that fire-gazers the world wide look forward to the date. This event can be considered a mini-eclipse since Venus will transit across the solar plexus of the Sun. This is a vast energetic network on the face of the Sun from which the chakra term of the same name has evolved. This passage will thus cast a shadow on the Sun's bright disc and presents an occurrence which offers an opportunity to learn much about our sister planet, whose luminescence has earned her the title of both the Evening Star and the Morning Star. The atmosphere of Venus was actually discovered during one of the 18th century transits.

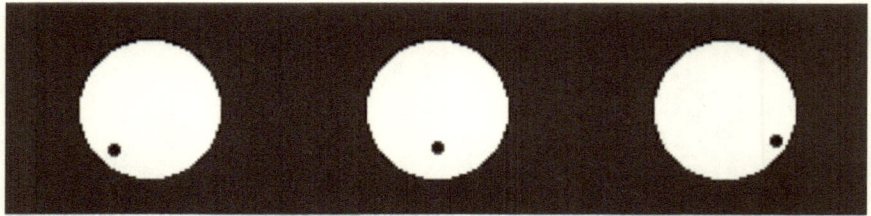

Figure 1. Venus in front of the Sun
From Fred Espenak's NASA web site

NASA breaks the transit into four "contacts." The first contact, ingress, begins when the disk of the planet is externally tangent to the Sun, and the second when the disk is internally tangent with the Sun. During the transit, Venus crosses the sun at a rate of approximately four arc minutes per hour. The entire event lasts a little over six hours. Contact three occurs when the planet's limb is internally tangent with the Sun, and the last contact occurs when the planet's limb is once again externally tangent to the Sun. The first two contacts are defined as the ingress phase, and the last two are known as egress.

Few transits have ever been recorded by astronomers, with the most recent transit of Venus occurring in 1884. Only one of five such events has ever been available in terms of technology so that it could be actually watched by humanity. Looking back through the corridors of time, one can chart the transits since the discovery of the telescope as follows.

June	December
1631	
	1639
1761-1769	
	1882
1874	
	2117-2125
2004-2012	

Table 1. Transits of Venus 2004, June 8

It should be noted that the transits of Venus occur during the months of December and June because it is during this time that the orbital nodes of Venus pass across the sun. As seen from our viewing platform, Earth, it is only transits of the inner planets that can ever be viewed.

It is fair to say that transits of Venus are among the rarest of predictable astronomical/astrological phenomena. They occur in a pattern that repeats every 243 years, with pairs of transits eight years apart separated by long gaps of 121.5 years and 105.5 years.

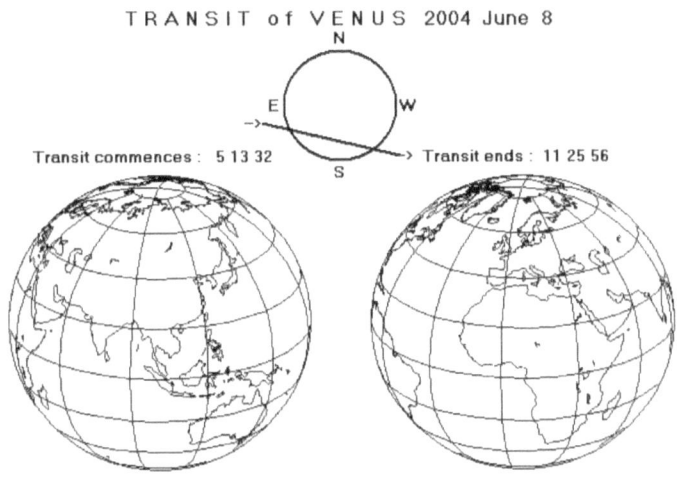

Figure 2. From The Royal Australian Astrological Society

The 2004 transit was visible from Europe, Africa except western parts, the Middle East, and most of Asia except eastern parts. Even though the transit was in progress, it was not visible because of sunset in Australia, Indonesia, Philippines, easternmost China, and Southeast Asia. The Sun was rising when the transit was in progress for the lucky viewers in western Africa, eastern North America, the Caribbean, and most of South America. Those who lived in western North America, Hawaii, New Zealand, southern Chile or Argentina were not be able to view the 2004 transit.

An interesting point can be made. There is always the possibility that the transits of the Earth can be viewed from the outer planets and perhaps, if life exits out there, some creature has probably noted our transits just as we observe those of Venus. Someday it is likely that an astronaut on Mars will view the Earth, as well as Venus and

Mercury, making predictable transits across the Sun. He may wonder at the beauty and perhaps draw conclusions about Venus, the Sun and our own planet, earth.

One thing is clear—today we no longer have to position ourselves in the correct place on the planet to glean information regarding this event because radar is used to monitor the process. We no longer seek to measure the astronomical unit, the distance between the Sun and the Earth, about 93 million miles, a task that tantalized and often baffled many world-famous scientists including Galileo, Hell, Copernicus, Kepler, and Brahe.

That work was largely accomplished in the 18th century when astronomers were sent around the globe to glean information. As an offshoot of these expeditions, many places were discovered— Australia, the Cook Islands, and Antarctica, to name a few. Countries at war cooperated to set up viewing stations around the world, so great was the over-riding interest in the 18th century search for valid data. It was presumed that if calculations could be taken from many positions around the globe, it would help astronomers calculate the distance to the sun. The determination to glean good scientific data led to an unprecedented period of international cooperation. Not only was astronomical data willingly shared, but also geographical and navigational information. So great was the crescendo of information that it spurred Captain Cook to further exploration not only in the interim between the transits, but also for at least ten years afterward. Unfortunately the 1761 transits yielded poor data and this only intensified the search for valid data. Fortunately, the second transit in 1769 yielded much better information as astronomers were edging toward defining solar parallax.

By the time the 19th century transits were due, the United States was not to be left behind. S.J. Dick (USNO) in a presentation, *"The American Transit of Venus Expeditions of 1882, Including San Antonio"* describes how the eight American expeditions were organized in 1874 by the Transit of Venus Commission. Simon Newcomb served as the Secretary.

The U.S. Congress appropriated funds totaling an astounding $177,000 for the expeditions. Although Newcomb considered the result of the 1874 observations disappointing due to inherent difficulties in the method, at the urging of the Naval Observatory astronomer, William Harkness, in 1882, Congress once again appropriated some $10,000 for improving the instruments, and $75,000 for sending eight more expeditions. Despite reservations about the method, Newcomb led an expedition to South Africa and among the four northern stations, Asaph Hall led the expedition to San Antonio, Texas. (American Astronimical Society, 187th AAS Meeting, #35.02; *Bulletin of the American Astronomical Society*, Vol 27, p.1331)

Harkness apparently spent much of his career studying the resulting data. His work culminated in his famous *The Elements of the Four Inner Planets and the Fundamental Constants of Astronomy*.

What scientists learned from the 18th and 19th century transits was a method of determining solar parallax. Parallax is a method of measuring the shift in an object's location by viewing it alternately from different vantage points. The understanding of measuring very distant objects by triangulation had proved a daunting challenge and was finally resolved when Stellar Parallax was discovered in the 1830's. Once this critical information had been digested, humanity's view of our universe was changed forever. Information of this magnitude gave insight into the scale of our solar system. In a broader sense, all the furor over the 19th century transits led to debates that nurtured what was to become modern astrophysics. No less an accomplishment was the eventual improvement in photographic techniques, which in the 20th century produced magnificent pictures not only of our own solar system, but also of outer space

Prior to the 19th century transits, great and competitive expeditions had been organized to observe the Transits of Venus. 18th century findings led astronomers to realize that that their data gave them a first good value for the Sun's distance. Edmund Haley, who discovered the now famous comet, was one of the first to determine that the transits of Venus could be used to calculate the Sun's distance. Although he never succeeded in pinning down the constant, the direction of his thinking stimulated others to proceed and succeed.
In 1631, the French astronomer, Gassendi, successfully observed a transit of Mercury, an event that occurs several times during a century. Buoyed by his initial success, he sought to observe a transit of Venus, but, unfortunately, he failed—the transit was simply not visible in Europe at the time.

Locational problems plus a whole host of troubles plagued the early fire-gazers. Obscuring clouds, ship-wrecks en route to viewing locations, fortunes plundered in one's absence, fever epidemics—such were some of the challenges faced by the 18th century scientists. A number of scientific voyages were commissioned, but although many resulted in disappointment, the overall navigational data encouraged further exploration of the Pacific area.

Today, viewing a transit is simpler. To determine whether one can see a transit of Venus from their home town, one must simply calculate the Sun's altitude and azimuth during each phase of the transit using information tabulated in the *Six Millennium Catalog of Venus Transits* online. Credit should go to Jeremiah Horrock. It was he who predicted and documented the first actual observation of a Transit of Venus in 1639, a feat that has earned him the title of the father of British Astrophysics. But bad luck followed him as Jeremiah died at the age of 22, a day before he was to discuss the event with the only other known person to have seen it.

Chapter Two

PROPHECY OR LUNACY?

The prophet is a fool, the man of spirit is mad.
Hosea 9:7

Already, there is a plethora of data available for the 21st century transits. The NASA web site, a real gem, gives detailed technical information, but it is not in the nature of NASA's purpose to assign meaning. What we know about the Universe and how we know it is at the core of both astronomy and astrology. Astronomers have used methods of triangulation, pendulum clocks, and telescopes for observing the activity of the heavens. Astrologers also use the motions of the planets and stars, as well as the angle of aspect—the relational positions that stars, planets, and sometimes asteroids make—in order to interpret the significance in the heavens. Astrologers consider the heavens and all that moves and evolves within the Cosmos as one vast creation in the making. We interpret the hints that nature provides us in order to re-awaken our soul purpose as well as to make sense out of our daily lives. We examine the planets as symbols of some facet of our manifested life. Sometimes as we scrutinize the layout of the horoscope, contradictions occur and it is the task of the astrologer to synthesize the apparent contradictions into some meaningful statement. There was a time not too long ago when astrologers and astronomers were one and the same. Even when formalized science came into being, astrology still had a close relationship with the science of astronomy.

The difference in their perspectives lies in the fact that astrologers ask a question: What is the *significance* of the events in the sky? Might there not be some symbolic correlation to human behavior?

When Venus makes its rare transit across the face of the sun, a question quite naturally arises: might these transits—brief though they be, lasting only about six hours in duration—hold some significance for humanity? Because Venus is a planet of love and money, the transit of this century suggests the possibility of reordering our basic priorities about what we love and what we value. The undeniable tension in the world between rich nations and poor ones calls for a reappraisal of wealth and possessions. Venus can teach us how to love and appreciate what we have as well as a concern for others. In this sense, the transit could be a kind of foreshadowing leading to a new morality—what humans owe to one another in a very basic sense.

The first transit of the 21st century set occurred on June 8, 2004, having begun at approximately 5:00 AM GMT. The process continued for around six hours. The following transit will occur on June 6, 2012, at approximately 5 PM EST and will last until around midnight. This transit will be discussed in a later chapter of the book.

The first of the two 21st century transits could be described as the Great Rehearsal, implying a time in which we may rethink the challenges and find solutions that will be influenced by the incoming energy. Presumably, there will be some kind of resolution with the second transit of the set. There is every indication that a struggle is ahead. We can expect an ever greater influence of Venus over the concrete mind of individuals and of groups. There will be deep seated urge for greater global harmony, greater ecological harmony and especially a more just and compassionate social harmony.

The 2012 event had been anticipated and described by the Mayans as the completion of the cycle of Creation. Does this mean the end of time, as some have predicted, or is it simply the completion of the first phase of Creation and the arrival of a new age? It may indicate a transition period of some kind, during which humanity will be tested.

Venus plays an important role in the Mayan calendar. The Dresden Codex, the best of the surviving Mayan treatises, marks the beginning of the Great Cycle as August 11, 3114 BCE. This date was known to the Mayans as the Birth of Venus. Actually, this may be the date that Venus was formally discovered by some long ago Mayan astrologer and it is included in the notations of ancient fire-gazers.

Why was Venus so important to the ancients? Many ancient people used Venus as a major point for calculating the calendar. The

Mayans believed that there had been several creations prior to the present one, and concluded that those creations and their resulting civilizations failed because they did not use a proper calendar. With that belief firmly entrenched in the Mayan psychology and religion, incredibly accurate measurements led to the creation of the Mayan calendar.

The following chart of the first Venus transit of the 21st century can be viewed as a reflection of our Sister Planet as she makes her rare passage across the sun. The light of the sun, on this day, as it sends the energy of the planet Venus, presents a rare and beautiful spectacle. While other planets are active, send their note, and will be considered, it should be remembered that it is the transit itself that sets the major note of the day. Let us see what Venus, our Cosmic Mirror, reflects to us.

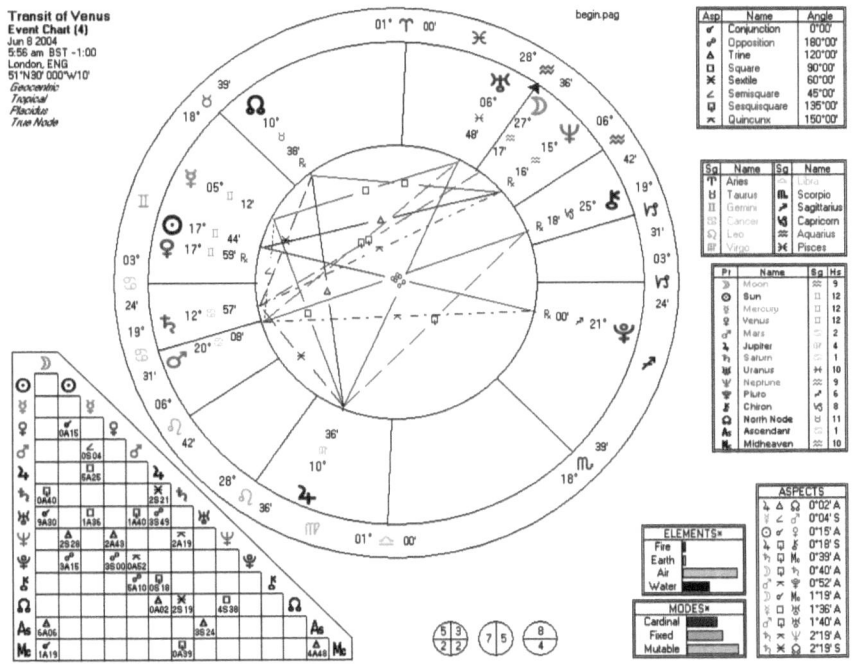

Figure 3. Chart of the Transit of Venus 2004

The Sun is at 18 degrees in the sign of Gemini (round off the 17degrees 59 minutes to 18). It has been said that Gemini forms a point of entry for the Cosmic energy of Sirius, so during this time

9

there will be the added influence of Sirian energy. Canis Major—the Dog Star, Sirius—has great meaning, particularly for the United States, because it conjuncts the Sun in the United States chart on July 4. For thirty-five days before and after the conjunction, Sirius is hidden by Solar glare, and it is during that time that the transit of Venus occurs during that period when Sirius is not seen. The star Sirius has long been exalted in history by its association with great empires. Sirius provides a clear channel for Cosmic energy to manifest its will. Situated below and to the East of Orion, the Great Man in the Sky, Sirius assumes the position of the faithful guard dog. The star is also known to be a symbol of power, the Watchman, and one who keeps guard over the abyss into incarnation.

With the 2004 (and the 2012) transit, the Sun resides in the sign of the messenger, Gemini. This position carries the energy of Mercury on the exoteric level and that of Venus on the esoteric. The conjunction with Venus adds to the potency of the Sun's illumination. Moreover, a Sun and Venus trine Neptune, represents an aspect that can elevate the consciousness to one of deep spirituality. Love-wisdom is strong with this combination. The Sun, Venus, Mercury conjunction squares Mars. Obstacles will be encountered and resolutions sought. Both Venus and Mercury are refining agencies. Struggles will surface and be confronted but courage and compassion will be present and available. Purification is in order. There will be a global struggle for supremacy. The struggle will entail soul-personality issues on all levels

It has been said that the higher octave of Neptune represents the heart of the Sun. With the added energy of Venus in the configuration we can expect an infusion of inclusive and creative love. On the mundane level, there is a tendency to fanatical devotion—be it to a cause or to an individual or to an idea. This kind of misapplication of energy is typically draining and nonproductive. Ways around impediments will be imagined, created and put, at first, to trial. The best of these practices will gradually take hold, fuse, and create a period of great social advancement. This will provide a much needed prelude to the Age of Aquarius which will take hold at the 22nd century transit of Venus.

While the analysis here is principally involved with the Venus transit, other aspects on June 8, 2004, resonate and either reinforce or impede the flow of Venusian light. With the Moon in Pisces, the principle of love-wisdom is once again reinforced. Mercury, as the agent of communication is strong in its own sign of Gemini, signaling an opportunity to detach from the past. The effort can be erratic as outworn paradigms are challenged. The Moon opposes Jupiter in Virgo, indicative of a time when Aquarian ideals appear to be frustrated and blocked, or when those blockages become public. The Occupy Wall Street movement is an example of this kind of pressure and activity to highlight needed change.

Public opinion can and will condition events. Opportunities to shape and blend social and political structures will emerge as humanity works with the prevailing energy to create something more harmonious. There will be difficulties as the old, outworn ideas—religious, political, business, educational—will not fade away without a struggle.

There is a natural association with Jupiter and Aquarius, and the Moon with Virgo. Those forms of human interchange which relate to the New Age may first surface, and appear blocked. But the Moon/Jupiter contact reveals a tangible underlying potential toward the blending of heart and soul. In other words, a way through presents itself. Venus, so prominent in this chart, can serve to throw light on obstacles as well as opportunities. The Great Rehearsal suggests that there is an opportunity to transmute basic, selfish desire to the more elevated phase of global aspiration and achievement.

With Uranus on the Ascendant we can definitely expect revolutionary ideas and change. Uranus in Pisces gives strength to group work, and expresses itself through love-wisdom. Group dynamics can be catalyzed by the Moon/Uranus wide conjunction. Jupiter square to Mercury bestows abundant mental potential—thinking in expansive ways or expanding communication. This communication can take an aggressive tone with the semi-square to Mars; nevertheless, there is plenty of energy and interest to quicken the process of dialog and conciliation.

Saturn sextile to Jupiter inclines humanity to compassion toward the needy. It can also trigger a display of art, especially music. Neptune trines both the Sun and Venus, thus bringing the potent energy of love-wisdom into the picture.

Mars in Cancer brings the ability and needed energy to nurture the masses while Saturn in Cancer lends structure, form, and the legitimacy of law. However, it can be a dangerous combination. There can be conflict with respect to selfish desire in opposition to the greater good. Pluto opposes both the Sun and Venus, re-emphasizing the opportunity to regenerate the past and to destroy old forms. With

Sun opposing Pluto, this break from the past can be violent, but the rest of the chart is so strong in the love-wisdom aspect that it does not seem to be the case. In any case violence can be ameliorated. Pluto in Capricorn will be with us for some time as humanity explores the right use of the will and aspiration. Pluto grounded in the sign of Capricorn suggests lessons that relate to all things material, to form itself. Pluto carries the First Ray and can lead to freedom from desire and form. It has power to release. It is highly possible that the necessary vehicles to achieve the activation of the will may surface at the 2012 transit.

The Venus transit chart clearly signals that a crisis of global significance respective to the affairs of culture and civilization is emerging from the depths to the stage of conscious awareness. Upheaval with themes of terrorism, poverty, and related frustrations are emphasized with Mars opposing Chiron in Capricorn. Chiron, the wounded Christ and the consciousness that it embodies, serves as a bridge between humanity and the higher worlds. Chiron in this position suggests the tension that can lead to violence—Mars—hence the wounding. The Mars position indicates that for the moment, violence will remain an issue.

On first glance, it seems obvious that this chart signals a wake-up call, and an opportunity to re-examine values. Human behavior is not determined by the stars, but presents us with an opportunity to respond to the higher energy that significant planetary events offer.

June 8, clearly, was one of those exceptional times in the history of the planet. The 2004 transit was best viewed from Eastern Europe and Western Asia, and was briefly visible in Europe and the Eastern coast of the United States. This may just be the little, but concentrated spurt of energy that is needed to further refine customs, behavior, and systems specifically but not exclusively in those parts of the globe. As for the emphasis on Eastern Europe, it is no doubt that the European Union (EU) has taken gigantic steps forward in the last few years, with its bold efforts to create a united economic bloc and a single monetary system—the euro. From a first glance at the chart, it would appear that this monetary system will flourish, and will continue to do so; the EU will be rewarded in its challenging attempt to create a freer, more flowing and open system, one which will greatly benefit the countries of Eastern Europe. Jupiter opposes Uranus, so prosperity will be associated with innovative approaches and fresh ideas. The question of debt and the restructure of financial systems must and will be addressed if further progress is to ensue. The theme of debt looms as a major global issue. It will be addressed and probably solved in the long years following the second transit in 2012. This theme may continue for many years into the 22nd century.

The initial effect of the euro on the United States and most specifically on the U.S. dollar will be discussed in Chapter Four. A point must be made—any consideration of the euro, the dollar, and any currency must be evaluated in light of the rising crisis of debt—a global issue. This is a situation which will be greatly impacted by the energy of Venus. We can expect better and clearer analysis from various points of view in regard to all things dealing with money, debt and credit.

The euro has served to tease Europe away from rigid nationalism and the long history of war into a vision of a unified continent. A single currency can and should lead to greater coherency and beyond that to prosperity. But more than a single currency is necessary to bring about the desired goal of a truly cooperative system. On the negative side, the specter of debt and all that it implies must be factored into that mix. Germany emerges as the economic giant that holds the cards for the continent. The European Union is not merely a useful tool for achieving national goals. Rather, it is an alternative to nationalism and the horrors that nationalism has brought to Europe. The European union is a vision of a unified continent drawn together in a common enterprise—prosperity—that abolishes the dangers of a European war, creates a cooperative economic project and, least discussed but not trivial, returns Europe to its rightful place at the heart of the international political system.

For the generation of leadership born just after World War II who came to political maturity in the last 20 years, the European project was an ideological given and an institutional reality. These leaders formed an international web of European leaders who for the most part all shared this vision. This leadership extended beyond the political sphere: Most European elites were committed to Europe (there were, of course, exceptions). Today, the future of Europe is in the hands of Germany. This is because Germany is the only county with the financial power to stave off the tremendous debt incurred by the neighbors to the south.

A look to the East reminds us that China with its gigantic potential as a creator of goods and services will also feel the infusion of energy. The transit goes over the 8th house of resources. Will it be simply a renewed interest in money for money's sake, with an exploitation of the poor so that the few may enjoy an abundance of luxury? Or can it mean something more? China has showed remarkable if not ruthless creativity in creating jobs, and improving basic living requirements since its birth. The citizens of the People's Republic of China have universal health care, but not always to the standards that countries in the West would necessarily salute. This system will be challenged significantly as more and more people flock to the big cities in search of the "good life" that globalization seems to promise. Nevertheless, after years of deprivation, one must conclude that progress has been made. With such a great proportion of the world's population living in this vast nation, fresh impetus to learn and see will necessarily emerge. Neptune has always played a significant role in China with its long tradition of spirituality, be it Buddhism, Taoism, or Confucianism. The Western concept of the ego is a far cry from the Asian association with the group as an entity. It is likely that this first transit will shift China even more so to center stage as a world power.

Sabian symbols, a set of sayings associated with particular degrees of the horoscope, are frequently employed to glean further information regarding an astrological event. The Sabian symbol for 18 degrees Gemini is "Two Chinese men converse in their native language in an American City." This suggests people who are entering into a new realm of activity where they may feel alienated and unsure of themselves. Yet if those people have something to offer, it is important that the new information be communicated. We cannot overlook the fact that China has become the largest creditor of the United States, a fact that heralds even more and more interaction between the two nations. Debt and credit are closely aligned with Venus.

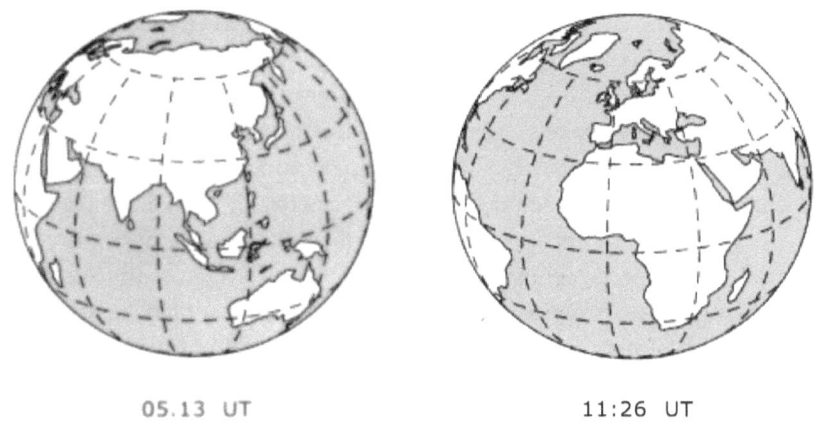

05.13 UT   11:26 UT

Figure 4. Visibility of the 2004 Transit

It is not possible to give one single interpretation for the Venus transit on a global scale because cultures differ and perspectives will naturally vary. This analysis is strongly colored by Western cultural and astrological perspective. That being said, and with the transit hitting Asia first, we must agree that the event offers an invigorating—even audacious—opportunity for the East.

Great strides have been made to eliminate global poverty, with China taking the lead since 1980 in reducing the poverty level for those living on $1.00 a day, increasing their income to $2.00 a day. This in a relatively short time marks a major global achievement.

But the larger question of poverty on a global scale remains. So does its relationship to terrorism. Poverty nurtures terrorism. That is not to say that there is a quid pro quo relationship. Poverty sets the stage for terrorism because it creates humiliation and stifles hope. People whose aspirations are so crushed join others, the alienated poor—in the Middle East, Indo-Eurasia, and the Philippines—who provide a spawning ground for terrorists. Add to that mixture the element of fanaticism and the situation becomes explosive. The UN has set the Millennium Development Goals, see the Appendix) as an opportunity to share the bounties of the Earth in a more just and equitable way. Abundance, overflowing bounty, such is the nature of Venus. At the very least, Venus implies an opportunity for humanity to learn more about right human relationships as well as the laws of the soul. Venus brings the "light of mind" which ties into right human relationships. Venus also reveals the Laws of the Soul.

If the sweetness of the Venus promise is to be fulfilled, rich nations will need to take a fresh look at their trade policies. This could bring clear definition to the Mars/Saturn contact in the above chart and channel the luminescent streams of energy in a positive way.

Within this setting, free will plays its part—humanity may respond, react, or overreact to these energies. That means we might take action in a positive way, we might fail to take action, or we might overreact. The latter will lead to war and even more violence. All three possibilities acting out on a world stage simultaneously suggest turmoil tinged with hope.

A World Bank study of 2003 indicates that the reduction of tariffs and agricultural subsidies by rich nations could raise global income in poor nations by $500 billion by the year 2015. This date will be only two years after the second Venus transit of this century. Unfortunately, that particular opportunity was declined at the 2003 meeting of the World Trade Organization. Nevertheless, the theme of opening markets will be a recurring one since such a maneuver could be a tonic to many nations—including Islamic ones—and bring more nations into the interdependent web of global trade and ultimately a more democratic way of life.

From this orientation to the 2004 Transit, we might conclude that there is a special meaning for Western Europe and the United States. These areas are certainly the main corridors of wealth and power on the planet—at least for the moment. It is in the West that the greatest accumulation of riches and monetary wealth respective to material resources lie. These concentrated areas will receive a significant burst of energy from the transit congruent with the restitution of good-will and elevation of right human relations. Certainly, there is a potential for such a transformation.

All this activity occurs under the influence of Gemini. This manifestation of energy in these corridors of power—the U.S. and the United Kingdom—is intensified by two facts. One important fact is that the Earth is the hierarchical ruler of Gemini and the other is that Venus is the esoteric ruler. With Venus and Gemini influence at play, planetary activity intensifies with all that goes on globally.

This can lead to the unfoldment upon our
planet of the consciousness of universality—to
which the word "Hierarchy" is the key.
Esoteric Astrology, Alice Bailey, p. 361.

On a larger scale, humanity itself is being challenged in terms
of what it means to be a civilization—a global community. A burst of
energy from Venus cannot be ignored. More than any other planet,
Venus is associated with riches, prosperity, and love. She has sometimes
been called the "Bringer of Peace." On that note, one might consider
the greatest resource available today to be our capacity to love, which
at the very least can be expressed as good-will.

As we look at the 2004 chart, it must be remembered that
the transits of Venus occur in the youthful, friendly sign of Gemini.
The Sun represents Will Power. Eclipsed by the Goddess of love and
beauty, as well as compassionate love of others, we can assume that
on a personal level this period will be a good time for lovers. On a
larger scale, aggressive egos will have a tendency to give way to peace
and unity. The alignment of Neptune at 15 degrees of Aquarius is
trine to the Sun and Venus at 17 degrees Gemini, thus adding depth.
Neptune is often considered to be the higher octave of Venus, where
Venus is the personal expression of love and Neptune is the universal,
unattached embodiment of Love. We can expect a real Aquarian zap!

The planetary alignment of Venus and Neptune suggests one
of the most refined contacts possible. It is highly likely that for the
mass public this contact can result in a refreshing surge of beauty, not
only through the finer arts but also in film and music. There may be a
resurgence of what, for lack of better words, we call good taste. The
Venus-Neptune connection, that of a trine or 120 degrees, is often
considered to be active, energetic, positive, and idealistic. The trine is
associated with creativity, idealism and higher mind. This configuration
can apply to both individual or to group horoscopes. It is an aspect
that brings vitality, strength, and ease.

As noted, Neptune is of special interest in this transit.
According to the ancient tradition of the Vedas, Varuna, a great Deva
or Angel, represents the astral body of our Planetary Logos, that great

Being that governs our sphere of life. Were we to add up all the lives and the attending consciousness of Earth's astral plane, Varuna is the result—and Varuna is a direct emanation of Neptune. An infusion of joy, beauty, and peace within the collective consciousness is bound to affect the realization that there are no individual problems. This kind of planetary awareness can only come about as we accept that we are all united in One Human Family. Neptune will nourish that perspective. As the great illusion of separation melts away, we can find meaningful ways to apply the lesson of unity and love to the world stage.

The deplorable working conditions of many of those who strive globally on a daily basis making our shoes, sewing our clothing, and picking our food will eventually be radically transformed. Without clean water, proper sewage, fresh water, and most of all hope, these disenfranchised people toil desperately meeting the needs of a hungry market place. Pain, sadness, disease, humiliation, and despair—all of these debilitating emotions are directly related to our being focused on a tiny portion of the picture. The Venus transit, though brief, will illuminate the world stage. With light and the ability to see more clearly, the question of inequality, harsh working conditions and child labor stand out as blights on the world stage. Prevailing business practices will be questioned, studied and revamped. With more light, how can we continue to enjoy the fruits of the labors of those so disenfranchised? With more light, how can we ignore the devastating impact on the planet? The energy from the transit will continue to shed light, realign our thinking and stimulate a more compassionate world view.

There will be a renewed effort in the pursuit of peace, but this effort will not be easy; a retrograde Venus can indicate war. An additional problem lies in the fact that Neptune can bring illusion to the scene. If conditions are not in place for a valid peace, no peace will occur. Misunderstanding and miscalculating the nature of the Neptune contact could lead to the attempt to make peace with an enemy who does not have the slightest interest in making peace. A robust sense of realism needs to be injected into any workable plan. We have to look at the whole context.

Certainly the whole concept of war, and how and when and how it is waged, is in order. With Venus retrograde, a serious public debate on the issue could highlight issues and clarify options.

> As man manages to rape the earth in order to demonstrate his power and intensify his pleasure and his sense of proud mastery, conflicts and disruptive processes are inevitably initiated.
>
> Astrological Mandala p. 94

Venus carries the energy to allow us to see the issues revolving around peace or war on a deep emotional-cultural level, especially with Neptune activated—more on that in a moment.

Saturn in Cancer suggests a return to real conservatism, not firebrand neoconservatism with its current rhetoric of frenzied patriotism, religious rhetoric and underlying violence. If the power inherent in world trade continues to exploit the poor, the obvious reaction will be further terrorism. But as Saturn creates structures sensitive to the times, powerful and needed changes can be put in place. The days of saving ourselves with force are over, although this will be a hard-learned lesson. Neptune makes that clear. Social justice and preservation of the land that Venus cast her eyes upon are at the crux of the matter. Venus retrograde presents a threshold of transformation.

Let us consider the Venus transit as it relates to the chart of the United States.

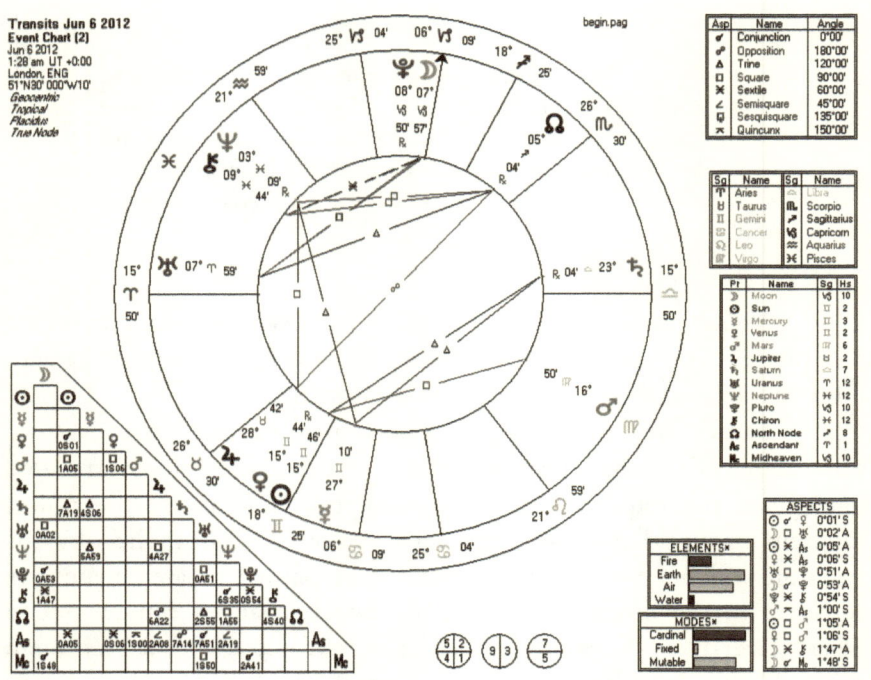

Figure 5. Chart of the United States and the Transit of Venus

This chart reveals important ramifications in the global political/economic sphere—Neptune conjuncts the Midheaven with Moon/Uranus in the 10th House. There will be unusual change and it will be public. It is clear that the United States has assumed an enormous global burden with Saturn conjunct the Sun/Jupiter/Venus constellation in the Second House. Transiting Mercury reinforces the notion of innovation of self-image as well as in communication style. Mercury conjuncts Uranus, an indication of revolutionary thought patterns. The United States must not be boxed into a position where it has to choose between the exploitative pattern of Ancient Rome and the overstretched tension of Victorian Britain. The Venus/Sun transit goes over natal Mars, a paradoxical situation that can bring light to violence as well as to the assertion of policy. The United States is attempting to fight terrorism and rogue states—transiting Chiron conjuncts Pluto, an interesting predicament. From one angle it can be an intense wounding of fixed institutions, of ecospheres—on the other hand an opportunity to bring in a regenerative will. Most important to this discussion is the Venus transit going over, as it does, the First House of the United States in this chart. The potential for illumination is great and in the direction of soul evolution. The trine to Neptune in the Tenth House can trigger a deep seated urge to dissolve obstacles to unity. The melting away of such barriers can, of course, lead to goodwill but the Pluto/Chiron conjunction will not be easy and it must be addressed.

Jupiter the carrier of love, wisdom and expansion squares Saturn. Jupiter can bring both cohesiveness and the energy of the heart into play. Jupiter likes to blend and with Saturn in play there can be a dynamic fusing process. Together they can bring about new parameters, new structures for expression. This partnership can bring attention to previous creative expressions (5th house Saturn) and unless there is strong influence from the Soul and higher mind, old and crystallized thought forms will hang on. This may cause some delays in local transformations but the Venus flow is so powerful that we can be optimistic. New channels of expression will surface. Moreover, the Venus transit in the First House emphasizes the optimistic spirit of The United States in its youthful, perhaps naïve drive to spread capitalism and democracy as "the way." Some sort of fusion is suggested with the Saturn/Jupiter connection. Above all else, Gemini is about relationship. It seeks to communicate on every level. What better vehicle than democracy to serve the basic impulse of Gemini? Venus as the esoteric ruler of Gemini carries the urge to unite and blend using the inflowing energy to create harmony out of conflict. The important question is how the United States will use its power in the future and how world opinion will react to it.

Fuel efficiency and the success of energy conservation (Saturn transiting the Sun in Cancer) loom as major considerations. A new Manhattan project is in order—one that will research and develop new sources of energy—and it is probable that science will flourish given the Venus influence. Moreover, Saturn can provide energy, balance and structure. One outcome can be the development of advanced technology to provide energy for world consumption. The chart is hopeful with Jupiter on the Descendant. This is where an evaluation of our deepest concerns comes into play. That is not to say that there is not tension. There is! Venus and Pluto are in opposition in the transit. Venus and Pluto make things happen on earth. The Trine to Neptune brings in love-wisdom and sheds light on the natal Moon in the United States Chart. The past can and must be transformed.

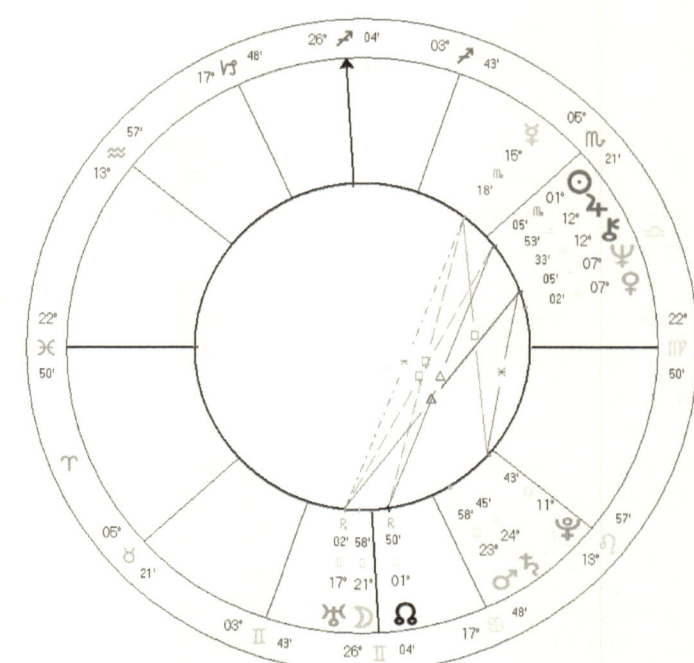

United Nations
Event Chart [2]
Oct 24 1945
4:39:30 pm EWT +4:00
Washington, DC
38°N53'42" 077°W02'12"
Geocentric
Tropical
Placidus
True Node
Rating: A

Figure 6. Chart of the United Nations

The 2004 transit affects the United Nations in the area of mental activity. Venus exactly conjuncts Uranus in the Third House— the area in the chart that speaks to communication. Process and the search for knowledge will be stimulated with this aspect. Additionally, Venus is the esoteric ruler of the Third House so that this placement can significantly influence relationships in the direction of soul expression. The transit will bring an overall infusion of light to the group of member nations and with the conjunction to Uranus, we may see much needed changes to the way in which the Security Council is set up. There will literally be a shaking up of the past forms with new expressions of political exchange and insights. These changes should be positive, with a powerful blending of the ideas of the individuals with the group. Women are on the Ascendancy. Programs like the Women Peace Maker program based out of the Institute for Peace and Justice in San Diego, CA, have provided a myriad of examples of how women make a significant contribution in the areas of Peace Making and Peace Keeping. Stories of women's activities all over the globe reveal that women have made a significant contribution to understanding the structural roots of violence and as well as those actions that ameliorate negative conditions. Women have been leaders and will continue to be so as we move forward into a New Age of Awareness.

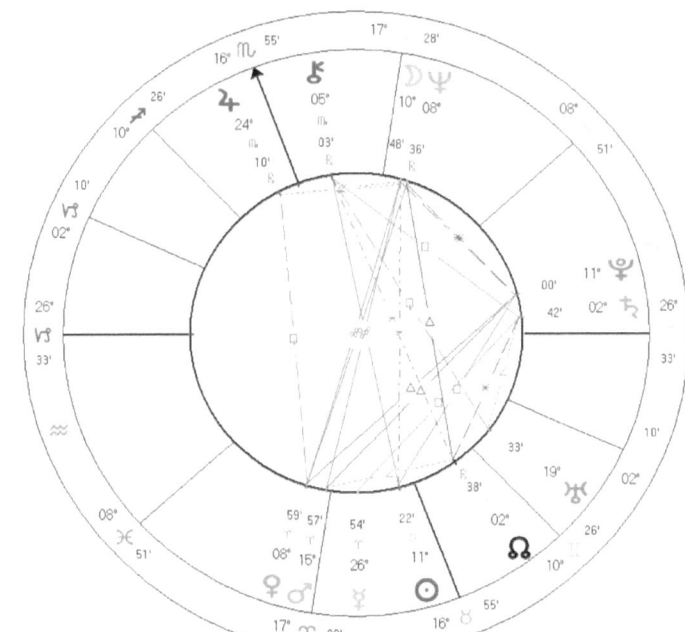

Figure 7. Chart of Japan

Chart of Japan and transit

Despite the fact that visibility of the 2004 transit coincides with sunset in Japan, nevertheless, that nation will feel the influence of the inflow of energy from the transit. With the Sun and Venus going over the natal Uranus there is a powerful and harmonious blending of the energies of change. Nationalism in Japan has been on the increase, intensifying with talk of remilitarization. The 2004 transit triggers change, but it need not necessarily be of an aggressive nature. Venus and Mars are conjunct in the natal chart, indicating the possibility of bringing reason, concrete mind, and the potency of Venus, to the passions and fears of a nation. Through this relationship of Venus and Mars, energy can be elevated, and as a result, intuition can be brought into the picture.

If Japan is sensitive to the Aquarian quality of Uranus it could draw closer to China, and together, recognizing the risk of an escalating arms race, form a security pact. This would require a radical change of thought as old memories are put to rest. China has already taken steps to bury historical divisions between the two countries, so this kind of shift is not out of the question.

Japan has taught the world much, and has instilled caution regarding the use of nuclear energy. The earthquake and resulting tidal wave that disabled the Fukushima plant in 2011 highlighted the inherent dangers of Nuclear energy, especially in earth quake and tidal wave prone areas. We know that an earthquake can occur anywhere and many plants are not safe. The country of Germany has decided to build no more nuclear plants and to put all attention of safer, renewable methods.

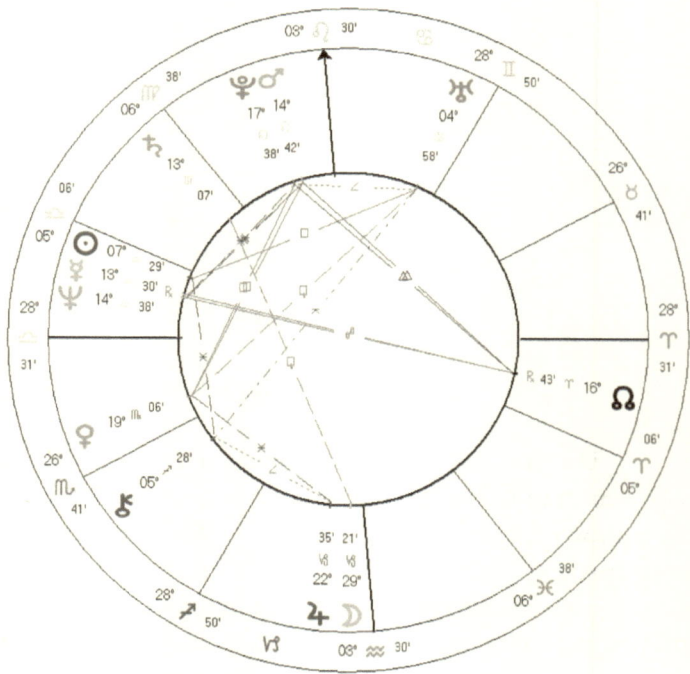

China
Event Chart (2)
Oct 1 1949
8:00 am AWST -8:00
Peiching, CHINA
39°N55' 116'E25'
Geocentric
Tropical
Placidus
True Node
Rating: A

Figure 8. Chart of China

Central and western China and much of Asia are also in a position to view the transit. Occurring as it does over the Eighth House, China's growing economy is bound to expand. Resources have been put to good use and China will transform material resources into a more refined culture for her people. Resources and Eight House matters remain a top priority. The world will continue to scramble for not only oil and gas, but also much needed mineral assets. The house placement contains the Scorpionic quality of tremendous focus—an obsession with economic growth. Foreign investment as well as technical assistance from abroad has made the boom possible, and this transit indicates that the momentum will be sustained and even increase.

China's emergence as an economic superpower has shorn up confidence, and for the first time in eons, China has stepped up to the plate and is acting as a regional economic leader. With new power comes a kind of international prominence that can prove unsettling to some of China's trading partners. It is essential that the United States find ways to deal effectively and fairly with China's growing success, with a mutual search for economic opportunity as a guiding principle. There will be continued pressure to get China to revalue the Yuan, a currency which is on the ascendancy. Right thinking and right motives must prevail.

Astrologers can suggest the trends and events. Thomas Szaasz once said that that if one talks to God he is praying, but if one says that God talks through you the world sees you as mad. Astrology is an attempt to translate the impulse of the universe in a meaningful way. The Great Architect provides a blueprint—we merely attempt to make sense of it. How those insights develop into plan and purpose, how they are used depends on humanity. The impulse toward higher spiritual development will be dealt with in greater detail in Chapter 7. On a practical level, perhaps what is called for is a new heroic race—one that will elevate, build though beauty, love and appreciate others, yet create a meaningful and rational peace.

Eventually, a sane human being must conclude that there must be more to living on this planet that merely satisfying our ambitions for comfort, pleasure, and personal security. The Venus/Neptune connection can touch deeply on feelings of personal worth and can spark a downpour of light from higher realms of consciousness. One thing is undeniable—the Earth is abundant in resources. How we wish to share those gifts will come to the foreground as Venus passes the first time and are deeply pondered with the second transit in 2012.

With transits as with eclipses, we look to the first event for the initiatory assertion of energy. Following along that line, the second event, in this case, 2012, will bring the culmination of the first impulse of 2004. It is interesting to note that the date, 2012, coincides with the Mayan prediction of the end of time. The Mayan calendar was based on cycles, and it is not clear what the exact cause of the end of time could be attributed to. Some authors suggest comet impacts, others magnetic reversals, or coronal flares. We can't, however, overlook a possible demise or at least a lessening to greed, selfishness, and separateness as a viable alternative.

One possible end in store for us is the end of the Age of Kali. The Vedas, the most ancient of Indian holy books, describe our current age as the Kali Yuga, a time of darkness and ignorance. Kalki, whose name signifies the destroyer of ignorance, is to make an appearance, and his coming ushers in a new age of peace, prosperity, and harmony. Research is on the side of optimism, for even great modern day Vedic scholars argue about the dates cited in the old calendars. Recently some experts put the onset of the Golden Age much closer to our own time than previously believed. Perhaps then, these current transits of Venus can be viewed in terms of Western astrology, as the end of the Age of Pisces, and herald the official entry into the Age of Aquarius, an unprecedented Age of Brotherhood. If we open our hearts and respond with right intention and goodwill, it can happen.

Christianity also prophesizes that extraordinary events will occur at the "end of days"—the resurrection of the dead and the Second Coming—to name two. Some Christian writers like Dostoevsky believed that it was the task of humanity to create an earthly paradise. Such speculative and transcendent visions of a state of total harmony may have an inkling of truth, if we consider the Neptune-Venus potential for synthesis. Even an approximation of such a dream calls for a major overhaul of culture and civilizations.

Human behavior, like astrology, follows certain laws that have a predictive value. Customs, values, and modalities of behavior grow from our culture. Social scientists of the 20th century tell us that culture is the most difficult phenomena to transcend—more difficult even than family values or national allegiance.

Every wave of humanity has had its flaws. It has been stated that the Lemurians had sexual excess to overcome. The greatest problem in Atlantis was excessive selfishness and the over-expression of the astral nature. But for us today, it has been said that the greatest sin of our age is that of separatism. Now, in the 21st century, comes a critical point—Neptune provides the opportunity for barriers to melt and Venus the opportunity to love. Our only limitations are those we cling to from the past. After all, a new age requires a redesign of a culture, a redesign that comes from within and without.

There is no doubt that as we move through the transits there is a general and pervasive sense that the inner processes are being stimulated. Times and energies are 'soul-size' as the unfolding takes place.

Chapter Three

A GEM IN THE HEAVENS

Figure 9. Botticelli, Birth of Venus

Shimmering, lustrous, and beautiful, our sister planet is similar in size, mass, density, and volume. Both planets were formed at about the same time and from the same condensed nebula. While astronomers agree that Venus is the second planet from the sun, many astrologers today would insist that it is the third. There is a growing belief among astrologers that a small planet, Vulcan, is the closest to the Sun. This belief was held by the 17th century mathematician, Le Verrier, after noting an irregularity in the motion of Mercury. Over time his findings of what came to be known as a ghost planet have been set aside, although Le Verrier held fast to his position.

To astrologers, the existence of the planet Vulcan makes sense for symbolic reasons. If Vulcan does exist and is followed by Mercury, then Venus is the third from the Sun—more on the symbolism of that later. A day on Venus is equivalent to 243 Earth days and is longer than its year, 225 days. Moreover, Venus has the peculiar trait of rotating from West to East. Without a doubt she is our sister planet though there are many differences. She has no oceans. A heavy mantle of carbon dioxide envelops her form, and her clouds are composed of toxic sulfuric acid. The surface pressure is 92 times greater that that of Earth, and her surface temperature is a scorching 482 degrees.

Yet at night, despite her mantle, she shimmers in beauty. Venus is the Roman goddess of love, beauty, pleasure, and wealth. The Greeks called her Aphrodite, and the Babylonians, Ishtar. The planet deserves her name because she was certainly the brightest light (with the exception of the sun and moon) known to the ancients. Venus has been and remains a familiar gem in the heavens since prehistoric times. It is believed that at one time she was known as two separate bodies, Lucifer, the Morning Star, and Hesperus, the Evening Star.

Civilization has gained much from association with the goddess—the ancient Venus calendar of the Mayans and Babylonians led to a coordination of sowing and harvesting with cycles commensurate with what was observed in nature. One ancient myth claims that Gaia, angry with her husband Uranus, cut off his genitals and threw them in the sea. The blend of his potency with the vast ocean produced Venus, who appeared against the background of sea-foam and stood free and radiant within a shell. Guided by prevailing winds and nature spirits, she appeared on the seashore. There she cast her eyes on the land, and desire was born. Her beauty was breathtaking. Undeniably, she provided a suitable form, within which the soul can reside and influence all she touches.

Myths are powerful, and give ways to organize experience into something useful. The past is obscured by sea spray, foam, and mist. The story of the birth of Venus gives us not only a glimpse into the intimate world of Woman, with it own very real psychological reality, but also insight into civilization, culture, and refinement. Temples have been built to Venus in the hope that some dazzle of blessing may come our way.

29

Present day astrologers may take two perspectives on Venus, or on any other planet for that matter: an exoteric view or an esoteric view. With respect to day-to-day living, exoteric astrology shows us how Venus relates to personal attractiveness, love, and romance. Her beauty is magnetic; she naturally draws to herself. She is about pleasure, and particularly pleasure with someone else. Myth has it that she was married to Vulcan, the god of fire, volcanoes, and the force. He was a master craftsman in brass and iron, and the manufacturer of arms, iron, and armor. Unfortunately, Vulcan was crippled and very ugly. Venus' beauty and yearning attracted a handsome lover, Mars (indicative of physical drive and sex). Vulcan learned of her infidelity and planned to trap the lovers. He was a blacksmith, so he forged a metal net and captured the lovers in the heat of their love-making. Having done so, Vulcan invited the rest of the gods to witness the offense. When the gods came and saw the two magnificent lovers joined in love and passion, they showed only respect and admiration. They laughed at the irate husband. It was Vulcan who was humiliated.

Venus deals with pleasure, not just physical pleasure but additionally the kind of enjoyment associated with possessions and the good life. There is always a sense of refinement that comes with anything related to Venus. Copper is the metal associated with Venus, and since early coins were made of it, no wonder we associate money/coinage with her. Additionally, she is associated with the South wind, and the qualities hot and moist. It was surely no accident that it was a friendly wind, the summer wind, that first blew Venus to the shores of materiality. There she could revel in the glory of our world.

In an individual chart we look to Venus by house placement and aspect to gain a better understanding of how a personality deals with love, money, and relationships. A person with a strong or well aspected Venus may have aesthetic temperament or talent, harmonious love, luxurious possessions, enjoy good food and drink, and have a beautiful home. Refinement and grace are key words. Venus is concerned with love, romance, and relationships—this can include marriages, partnerships, and friendships. The Venusian impulse is to spread happiness, joy, and truth.

Soul power includes three elements—the power of will, the power of attraction and the power of intelligence. All these elements will come into play with Venusian energy coming into the world stage. With will comes the ability to confront obstacles, with attraction comes the ability to love and to sense unity, and with love comes the ability to serve others.

Were it not for Galileo and the invention of the telescope, humanity would not have had the joy of witnessing Venus. Galilei Galileo was not only an astronomer but also an inventor, mathematician, and physicist. By a tedious method of trial and error he was able to teach himself the secret of lens-grinding. Intensely committed to a path of continual improvement, over time he produced a spy-glass. He then went on to improve the process until he was able to create even more powerful telescopes. For his work, he was awarded a prestigious professorship at the University of Padua in the Venetian Republic. Galileo's discoveries were paradigm shattering. He brought truth and light to bear on the subject of astronomy. Just one of his many discoveries—the fact that Venus, like the Moon, goes through phases—was enough to make a benchmark.

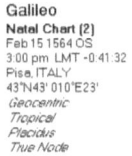

Galileo
Natal Chart (2)
Feb 15 1564 OS
3:00 pm LMT -0:41:32
Pisa, ITALY
43°N43' 010°E23'
Geocentric
Tropical
Placidus
True Node

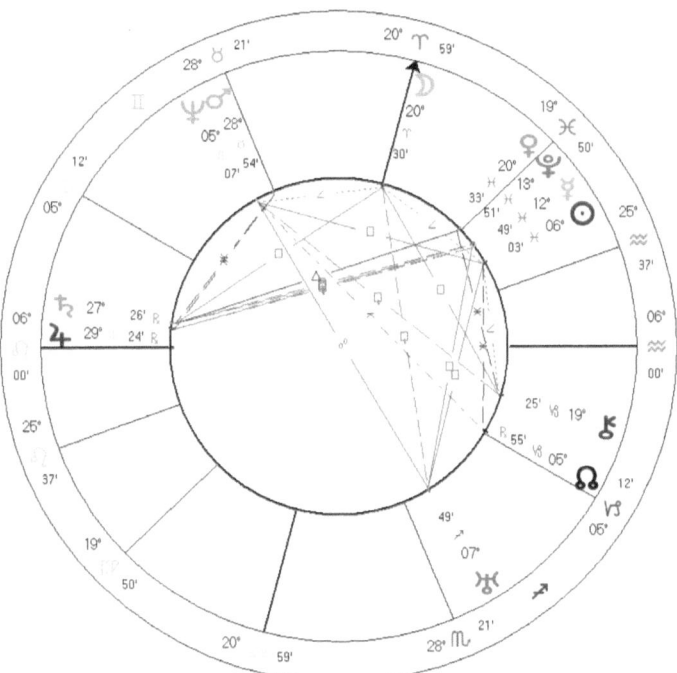

Figure 10. Chart of Galileo

With Jupiter on the Ascendant, Galileo has a royal chart. It brings good fortune overall. The Sabian symbol for the Ascendant is "The constellations in the sky." With his Sun in Pisces in the Eighth House, he could expect a life that would bring transformations. The Sabian symbol for his Sun sign is, "A cross lying across a rock." Galileo would challenge the orthodoxy of the Catholic Church, but despite everything, the Saturn/Jupiter conjunction bound him to the old, established religion. Sun/Uranus square provided leadership skills as well as a strong will. Not always tactful, such a person would remain an individual to the end. Galileo was bound to have a public life. The Jupiter influence on the Ascendant was re-emphasized with the Moon on the Midheaven. The Sabian symbol for the Midheaven is "a pugilist entering a ring." With the Moon in Aries, there is a life that reorients the emotional body to the requirement of the soul. The Leo Ascendant expresses the need and the ability to lead. The intense self-consciousness of Leo is poured into his scientific work so that others may see. Galileo's display of will is essentially linked to his spirituality with Pluto in Pisces. Moreover, Pluto conjunct Mercury points to highly developed mental powers. Venus trines Saturn—he was able to draw on the past discoveries and structure, bring that information forward, and improve upon it with new interpretations and inventions.

These improvements brought Galileo into a collision course with the Roman Catholic Church and eventually the inquisition. Galileo's discoveries were so convention-shattering that he was seen to be obstructive to the teachings and authority of the church. All the data that he found only served to reinforce the Copernican view that the Sun, and not the Earth, was the center of our system. Even though his opposition to the Ptolemaic system marked a high water mark in courageous scientific and philosophical thinking, it would bring him to an inevitable face-to-face encounter with the prevailing powers—the Catholic Church. At stake was who would explain the objective world. Who would define the parameters, the cycles, the forces? Would it be pragmatic and scientific? Would it be ecclesiastical? Galileo held fast to the use of mathematics, and objective research. He was sensitive to the Venus influence of scientific analysis and reason.

Galileo was tried in 1633 during which it is said he rose to his feet and uttered the now famous cry, "Nevertheless, it does move." To save his life, he recanted. Galileo spent his last years in seclusion under house arrest. He used the time well, writing and re-writing; he persisted in seeking out a way to reconcile scientific truth with church doctrine.

On the esoteric level, to gain insight into soul awareness and growth, we can look to Venus to grasp the progress that has been made by the soul in a particular incarnation. Venus is the form to which we look for the expression of a higher love, the Love Principle itself. Venus is exalted, expressed in the best possible way, in Pisces, the sign of the Christ, and has a naturally harmonious bond to Neptune. On the soul level, Neptune and Venus are extremely complementary, serving to blend their energies and help the personality become something more—a truly alive and harmonious being. If and when this state becomes stabilized, we have a Master. Such a one was the Buddha, such a one was the Christ. These humans, who became something more, display a level of development in which ordinary knowledge has been transformed into wisdom.

In esoteric astrology, the Moon signifies the lower nature as well as past history. The Sun signifies psyche, the soul, and the deepest intent of a particular life. We study the movement of the Sun around the ecliptic (the path the Sun makes through the signs). By studying transits and progressions in the natal chart we can determine probable outcomes for various situations. The Venus transit lends energy to soul-force because the energy of the Sun is coming through Venus for that short period and thus magnified.

Fascinating, is it not, to see how an individual chart can be subject to two quite different though complementary levels of interpretation? Similarly, with the 21st century transits of Venus, an interpretation can be made on two levels. With the Sun and Venus conjunct in Gemini, it is easy to see how one could interpret this to mean an intensification of interest in material objects, a fresh fascination with beauty, and a restoration of refinement. Similarly, Venus will shed light behind any relationship, not just the personal. Influenced by Neptune, a planet often called the higher octave of Venus, the opportunity will arise to reconcile opposites. Such a simple problem, "Can I have a good life and also support, at the very least, a decent, good, wholesome, Venusian standard of living for my fellow human beings?" This is just one example, one important query for the potential for reconciliation. The transit suggests the possibility of synthesis, a blending of polarities into a third beautiful harmonic, one that could be termed a more elevated outcome. This question of dealing with the poor on the planet may be the most important one of our lifetime.

The astrologer can look to an individual chart to see where the transit may have either or both an esoteric or exoteric impact, because the potential for synthesis lies within every individual's chart. Not to be overlooked is the group effect. A critical mass of inspired souls can have a magnificent effect on the whole of humanity.

The question might arise as to how exactly a Soul-centered chart might be approached. Squares, oppositions and conjunctions will demonstrate those areas in the life that need attention or that will be emphasized. This could take the form of rethinking one's orientation. Balance may be required. Areas of tension would require discrimination involving Soul centered rulers of houses in contrast to personality rulers. Similarly, the aspects of ease or trines need meticulous evaluation. A fluid flow of energy can be supposed in these cases. One can in this case, ponder ways to optimize the creative, lively stream that a trine can provide. An easy connection can be made to make better use of the individual's basic and Ray make up. In the latter case, where there are aspects of stress the proper use of these energies can avert a crisis or at the very least ameliorate one. This means crisis producing aspects and these can be consciously optimized for the good of the soul. When the subject has succeeded in aligning personality with a conscious control of the physical, emotional and mental nature, he will be able to bring the forces of all the planets and aspects into high level spiritual play.

As far as the planets are concerned, it is not a great leap to make the jump from the exoteric meaning of Venus as beauty, light, and wealth to a higher, soul-centered interpretation as suggested in this chapter. Intuition and imagination are imbedded in the aura of Venus. One can accomplish a great deal on one's own in terms of comparisons by simple analysis of the accumulated knowledge in exoteric astrology. Logical deductions can be drawn. Suggestions for further study of esoteric astrology are given at the back of the book.

Energetic flows and expressions are significantly heightened by the fact that the Sun in the case of the 21st century transits of Venus, is in the sign of Gemini. The Sun naturally brings the warmth of love and creative energy into play.

Figure 11. Glyph for Gemini

The glyph for Gemini reflects a state of duality. Gemini is the source of the concept of the two pillars, so familiar to the Masons. Gemini guards the mystery of duality and offers the initiate a way through. This is the force that produces changes needed and forms a point of entrance for the cosmic energy from the star Sirius. The glyph demonstrate two poles, two brothers—this suggests personality and soul. Gemini carries the secret of relationship.

Relating the sign Gemini to Venus, the esoteric ruler of the sign, we can see the real possibility of resolving the basic duality of life in form—spirit and matter. Gemini relates, communicates, and as we shall see in Chapter 6, manifests strongly through the Second Ray of love/wisdom. Venus carries the potential of transforming all the dualities of life, not just that of spirit and matter, though that certainly can be viewed as the mega-paradigm.

Relating the sign Gemini to Venus, the esoteric ruler of the sign, we can see the possibility of resolving the basic duality of life in form—spirit and matter. Gemini relates, communicates, and manifests strongly through the Second Ray of love/wisdom. Venus carries the potential of transforming all the dualities of life, not just that of spirit and matter, though that certainly can be viewed as the mega-paradigm. A look at the human scheme of evolution is useful to better grasp the process. From top down in the line up of the Human Composition we can envision the process of evolution. Spirit enters into matter. From bottom up we can see the process of return, or evolution.

The question of which is the third planet from the Sun remains. If Venus is, indeed, the third planet, counting Vulcan as the first, and Mercury as the second, the change in the lineup will affect the rhythm of the spheres. Some still count Earth as the third planet, but this system which adds Vulcan changes the corresponding distance as well as the numerological significance. It makes sense that Earth should be the fourth planet since the Fourth Ray of Harmony through conflict is associated with Humanity who inhabits the Earth. Harmony through conflict is the way of Humanity. It is how we learn, how we progress and how we move forward, attracted as we are by love. Love is progressive; it is the way of service and of sacrifice. Every time the returning spark encounters resistance, it struggles and finds a way. It is through will, love and intelligence that humanity will find its way home but no one can deny it is a struggle.

Chapter Four

THE EYE OF THE BULL

So we stood, alive in the river of light
Among the creatures of light,
Creatures of light
Ted Hughes

The myth tells us that Psyche, a mortal, was so renowned for her beauty that humans began to worship her, and instead of making pilgrimages to Aphrodite, the Greek version of Venus, people sought out the lovely princess. As the shrine of Aphrodite became empty and neglected, Aphrodite became so jealous of Psyche's beauty that she conspired to have her fall in love with an ugly man. But as chance and circumstance would have it, Aphrodite's own son, Eros, fell in love with Psyche. The rage of Aphrodite was not to be easily mitigated and it was only after several complicated turns and twists of fate that Eros and Psyche were wed. In the tumultuous love story, Psyche tastes of betrayal and is exposed to not only fame but also anxiety, doubt, and grief. It is said that the offspring of their marriage is pleasure. At the end of the story, Psyche finds herself in close proximity to Love at the top of a mountain. A similar Roman myth comes to us in the characters of Venus, Psyche, and Amor. The story of Psyche and her love has been a popular subject of art through the ages, and bears a particular relationship to the timing of the transits of Venus.

One thing to remember in the myth is that Psyche ultimately achieves immortality. It is the immortal soul that is interesting. The soul ultimately inhabits and animates the body. But the body is not what lives on. Similarly, the role of humanity is to be free. To express freely, released by beauty into the eternal light where truth is perceived—this is our destiny. Love is the way.

The astronomer, Alan Heirshfield, points out that the asteroid, Eros, Love, is simply an irregular chunk of rock. Its orbit around the sun carries it as close as 4 million miles away from the Earth. Only asteroids stray within triangulation range of the earth. (Triangulation provides a way for astronomers to measure distance.)

The soul has been of interest to philosophers down through the ages. The ancient Egyptians believed in a concept of a soul—an immortal component of man. Perhaps it was from the Egyptians that the Greeks adopted the concept for certainly Alexander the Great's stay in Egypt afforded him the opportunity to become acquainted with Egyptian mythology. They believed in a god-like essence that pervaded all life. In humans, this essence was believed to be immortal and hence elaborate funeral rites were developed to provide for continuity of the individual soul. Elaborate funerary rituals typified Egyptian culture and most likely influenced Greek thought.

Plato saw the soul as the ordering principle of the universe—separate and apart from materiality. It was soul that gave life to vegetative and animal existence. He believed that the soul gained knowledge of itself by reflection of its own acts.

The Greek word psyche and the Latin anima both connote life as the underlying concept. Some go further and say that the soul, when in the body, is the very source of life. We can, for purposes of astrology, conceptualize the soul as our connection to the Source of Creative Intelligence and Divine Will. The soul, in this context is a small individual unit of life that can be linked to great Transcendental Beingness.

The Greeks, Plato, and especially Aristotle linked soul to the final, efficient and formal cause of all bodily movement. Aristotle believed that the soul and reason were intimately connected. The crowning ability to reason is what makes man preeminent in the universal scheme of things on earth. Reason can be applied to real problems and solutions developed. Reason allows for the creation and appreciation of art. Aristotle believed that although man partakes of the lower order of things—plant and animal life—he maintains a distinct and unique primacy.

To have a soul gives certain functions. The Greek view held that the soul could be looked at in terms of the both irrational and the rational. They also held to a continuity of all life with the soul present to some degree in all forms. The higher the form, the greater the degree of rationality.

Within this interpretation, the soul was seen as the source of the livingness of the body as well as a source of inner meaning. A human then must be seen in terms of potentiality. Linked to reason man's soul can be taken as the essence of the whole living body. The soul is the inner meaning, an internal gyroscope.

An Esoteric view of man is congruent with Aristotle's paradigm. Esotericism, however, would include the mineral kingdom as part and parcel of man's composition. Esotericism also gives very discrete descriptions of the higher states available in the process of evolution. Psychic energy is described in terms of higher and lower. One who is infusing more and more soul light into daily life through sacrifice, discrimination, right speech, and self-forgetfulness is tuned into higher psychic energy and will eventually develop a certain continuity of consciousness.

This is not to say that those who have some degree of soul light do not have powers. It is possible to exhibit clairaudience, psychometric powers, imagination, emotional idealism while still in an early developmental stage. These characteristics can emerge and exhibit well before true illumination occurs. Higher psychic development includes activities that elevate others, create great works of art, and that are inclusive and highly intuitive. A clue to this type of illumined mastery is found in many archetypes including Hercules, Krishna and the Christ.

> Christ was and is the greatest Psychic ever produced by the human race. We are told that the coordination between His endocrine glands and etheric centers was unsurpassed. His physical, mental and intuitional senses were coordinated in such a way that He was in direct communication with the higher worlds in His physical brain consciousness. He demonstrated divinity through His humility through His divinity. That is how one can be a Son of God.
>
> Saraydarian, *The Psyche and Psychism.* p. 72

With infusion of soul light, development becomes certain. Powers emerge naturally. Intuition prevails. This kind of systematic development is often described as the path of the disciple.

> From darkness to Light
> From the Unreal to the Real
> From death to Immortality
> From Chaos to beauty.
> From "The Upanishads."

In "Discipleship and the New Age," Alice Bailey reminds us that before we can tread the path we "must become the path itself." She is, of course, speaking of the soul and its journey toward unity sometimes called the return home. The Soul, is a returning Spark which once formed a great body composed of millions and millions of cells or sparks. The soul fell or descended into matter (involution) and is, therefore, on a journey of return. This return to the light, we can call the path of evolution. Hints of this journey are mirrored in the astrological chart of any individual, group or nation and it is for this reason that both an exoteric and esoteric meaning are possible. The life as it is lived in the moment, has certain challenges. But in the larger scheme of things, there is another picture, one of evolution and enlightenment.

The simple model Bailey gives depicts components of soul— the three midway items depicted below. These midpoints present a path to an even greater unity within the whole. The chart begins with the highest point of spirituality and shows less and less evolved or integrated levels of Beingness. This downward movement is called the path of involution. This is the path into matter and into form.

All these levels make up what we know as a human being. The degree to which we are aware of the levels is key to one's personal and group evolution. The degree to which we integrate higher energies into daily life corresponds to soul infusion.

It is for this reason that the Transits of Venus present an opportunity as well as clues as how to best proceed.

The Eternal Spirit
Monad
Spiritual Triad
(Higher Self, Intuition and Spiritual Will)
Solar Angel
The Soul
(Higher Mind)
Angel of the Presence
Persona
(Physical body, desires and lower, concrete mind)
Dweller on the Threshold

Figure 12. The Human Composition

Bailey is suggesting that the soul is a kind of way station in the process of Ascension. To even begin the journey seriously, some level of integration must take place between the Higher Mind and the spiritual triad. The word *antakarana* is used to indicate the approach to unity—an even higher state of consciousness. This midway area of the soul is frequently referred to as the rainbow bridge.

A soul-centered individual, therefore, is one who has at least achieved some minimum level of integration, one that provides a platform for more advanced work. At this point one could profit from astrological insights that shed light on the kind of development that guides one to even <u>further</u> potentials of integration. Astrology, in this case, alerts the subject so that one may be better prepared to make the approach to even higher levels of consciousness. A study of one's chart can be extremely helpful to gain better understanding that can lead to facilitation of the process.

Astrology, in this case, can be used as a tool to alert the subject. In this way one may be better prepared to make the approach and eventually ascend to ever higher levels of consciousness. A study of one's chart can be extremely helpful to gain better understanding that can lead to an overall facilitation of the process. Strengths can be analyzed, weaknesses addressed. These same techniques can, of course, be applied to any individual's mundane life.

In exoteric, or mundane, day-to-day astrology, Venus represents relationships, how we connect with one another. As we look to soul-centered astrology, Venus energy plays still another part in the drama of our lives. Venus embodies the Law of Magnetism. Over time we learn what Psyche learned in the myth—entrance to heaven is through love. It is not unusual, therefore, for a greater tension to develop between what we desire and what we love as we consciously take the path of illumination. Old assumptions and habits will be challenged both individually and collectively.

According to the Tibetan Teacher, Djwhal Khul, a New Group of World Servers (NGWS) is now coming under the influence of Taurus, which is ruled by the illuminating energy of Venus. The NGWS has the potential to serve humanity in an Aquarian manner, congruent with the approaching New Age. Their presence may or may not be more visible as time passes but their effect will be notable.

The Eye of the Bull radiates the influence of the divine energy of Taurus as it brings illumination and the attainment of vision. The group can be taken collectively as the "bull, rushing forward upon a straight line with its one eye fixed upon the goal and beaming light." To what is the Bull rushing? Not self illumination, for it is assumed that it has already accomplished that. It is something on a larger, vaster scale—the goal of providing mankind with a center of light-giving force. This will serve as a magnet toward which humanity will be drawn. In this scenario, desire is transmuted into aspiration. Thus darkness gives way to light and illumination, and the Eye of the Bull opens. This can only mean the ajna center or third eye of intuitive knowing. The Bible states, "If thine eye be single, thy whole body shall be full of light."

With this opening comes the urge to overcome desire and the longing for liberation that the Buddha sought and attained. The eventual transmutation of desire into Love is accomplished, and the purpose of life is fulfilled. For this reason, the sign Taurus has a special tie to the Buddha, who achieved victory over desire and became an illumined Master.

The glyph of the sign of the Bull with its upturned horns and circle below is an appropriate symbol for this dynamic process. As man, the Bull of God, surges forward toward the eventual goal of release from desire, and thus leads to illumination. The soul emerges from bondage, and the two upturned horns serve as a protective mechanism for the forehead. Traditionally, this is where we find "the third eye." The third eye relates to higher level of consciousness and when it is open it signifies "Enlightenment." It can have an enlivening effect on all other chakras.

As we are focused on a dynamic process we can imagine going around the zodiac from the beginning point. We can summarize this journey in the following way—the Ram of Aries, signifying the starting point of life, has sought the illumination of Taurus and will eventually lead us to the Mountain Top of Glory in Capricorn. Here, still another transformation will occur, and the Bull eventually becomes the Unicorn. It is interesting to note that the Unicorn has only one horn—it is pointed straight ahead. We had begun the process of movement around the Zodiac with Aries whose horns encircle and point downward. With Taurus the horns encircle and point up. Much mystery in encoded herein.

Within this paradigm, Taurus, Venus, and Desire are interactive. They interconnect by means of magnetism. The higher mind will struggle to maintain clarity but there will be a struggle. Taurus pours the necessary energy to stimulate desire on our Earth, thus setting up that magnetic field wherein the Will of the Most High is operative. To be practical, Taurus forges the instruments of constructive living with the aid of Venus and Vulcan. There is also an opposite side to the coin because where there is creation, there is also destruction. Scorpio is the polar opposite of Taurus and this tension must be considered. Taurus forges the karmic, materialistic chains which bind. On the opposite side, Scorpio demands the challenge wherein all hidden vices and limitations will be brought into view. So it is that Venus draws to the light, and it is this light which brings about release. This stirring will eventually reveal the mystery of life. Within this interaction, Vulcan cannot be ignored. He controls the clamoring anvil-like processes of time and strikes the blow which shapes the metal into that which is desired. (*Esoteric Astrology* pp. 374-75.)

With Taurus comes a definite urge to express on the physical plane. As Venus transits the Sun—as viewed from the Earth secrets of divine purpose and planning can be glimpsed, if only briefly. Self-willed destructiveness will be challenged. In the face of this challenge:

> *Let us build our fortress, aspire and affirm*
> *the principles of Brotherhood.*
>
> *author*

This kind of a response can only serve to reinforce the spiritual will. We can learn to use our material and spiritual resources purposefully.

Figure 13. Image on U.S. Dollar Bill

When the United States was established, Masonic principles strongly influenced the Founding Fathers in their platform for social change. It is interesting to note that using a set of coordinates based on the Mayan concept of yearly cycles—the k'atun—the addition of the levels on the pyramid adds up to the year 2012. Is this a point of termination or a point of a major recapitulation? On the dollar bill the k'atuns are represented as tiny building blocks of the pyramid. This revelation has a point of convergence with prophecies which use that year as one of major transformation. The Masonic thinkers set out to build a new world order—Novus Ordo Seclorum—in the United States. This saying is clearly visible on the dollar bill. The eye is seen at the top of the pyramid. Underneath the figure the words of the Great Seal make a profound statement about the country and its founders.

Given this information, we should be prepared to address the sober challenge of creating a new financial institution. The present system is running out of time. New systems as well as fresh decision making creations are in order. All these things will work out after the 2012 transit. Light will be shed on what has developed into structural blindness. At first there will be the slow difficult process of creating alternative institutions. This will take time. There is no doubt that the present model will be greatly overhauled. Venus brings the ability to think and act sensibly as well as scientifically. Keen analysis as well as mastery of factual detail will come into play. Debt, investment, financial markets, government structures, interventions and the federal reserve (in the U.S.) will be rethought, redesigned and recreated. Opportunities for increased consciousness will pervade the entire system as it rises anew to meet the demands of a New Age.

Is this the end of capitalism as we know it? There is a good chance of that. There is also the possibility of some kind of hybrid influenced by the Chinese model—a blend of capitalism, American style and communal thinking. There will be experimentation and analysis along the way. But the new is coming.

The dollar will endure for the immediate future. It is the fate of the euro that is at stake in the immediate future as the European Union struggles with its debt problem. In the end, the dollar may be challenged by the Yuan. In any case, as the years wear on, a very different system will emerge.

Chapter Five

Vibrant Expectancy—
Setting The Stage For A Season Of Light

*When the last transit occurred the intellectual world was awakening
from the slumber of ages, and that wondrous scientific event which has led
to our present advanced knowledge was just beginning. What will be the
state of science when the next transit season arrives God only knows.*
—Harkness

There has been a sense of expectation, an increasing awareness that
we are living in times that truly are "Soul-size." In 2003 we began a
second order cycle of Mercury. Additionally, there have been two other
second order cycles—of Venus and of Mars. Second order cycles tend
to affect society as a whole, and have to do with group behavior and
opportunity. The present cycle of Mercury lasts for about 66 years and
will color the times in which we live. (Second order cycles occur when
the Sun conjoins the planets node, usually the north node. For the
inner planets this begins at the inferior conjunction.)

On August 25, 2003, Mars made the closest approach to the
Earth since approximately 70,000 years, a date which has been linked
to the fall of Atlantis. The proximity of Mars plus the added effect of
the Mars cycle must necessarily cause an experience of quickening as
we respond to its infusion of energy. The Mars cycle is associated with
planetary activity. Mars rules the kundalini, so putting it all together
we can conclude that there is a general warming and rising in the
system. Kundalini is that powerful and mysterious electric liquid that is
associated with living light. With the proper stimulation it is activated
and rises within the spinal column, bringing with its ascendancy an

expansion of consciousness. Stress can be experienced when the rising kundalini clears out blockages, but eventual enlightenment is the prize. On the planet, there has been some seismic activity—Arenal in Costa Rica erupted in September of 2003, Kilauea on the Island of Hawaii continues to spew forth molten lava, and the Ring of Fire in the Pacific rumbles with encouragement. More seismic activity is likely, as volcanic activity, magma flows, and plate tectonics respond to the increasing heat of the planetary core. With Pluto the transformer in Capricorn, earth changes are to be expected.

Mercury rules the active liquid of the kundalini. For a compelling articulation of the role of Mercury see *Mercury, the Divine Messenger (hidden cycles in momentous times)* by Malvin Artley.

The long progression of Pluto through Sagittarius brought to the foreground not only idealism and also the prevailing global fanaticism. This awareness necessitated a reevaluation of values up to and even to include the present time. A sense of global tension pervades the atmosphere. There is still time to integrate lessons learned from past years' Sagittarius/Pluto contact. Yes, idealism has led to fanaticism but is also allowed humanity to envision an opportunity to contact the reality behind form, to penetrate the hidden and veiled.

An interesting moment, and one that relates to the Venus 2004 transit, was the November 8, 2003, full moon in Scorpio. This event stands out as important because of the fact that it was an eclipse and eclipses are naturally more important than ordinary full moons. The Moon itself was in the sign of Taurus, and the eclipse occurred at 15 degrees. This fifteen degree point signals a point of great release in the fixed signs—the Bull, the Lion, the Eagle, the Man.

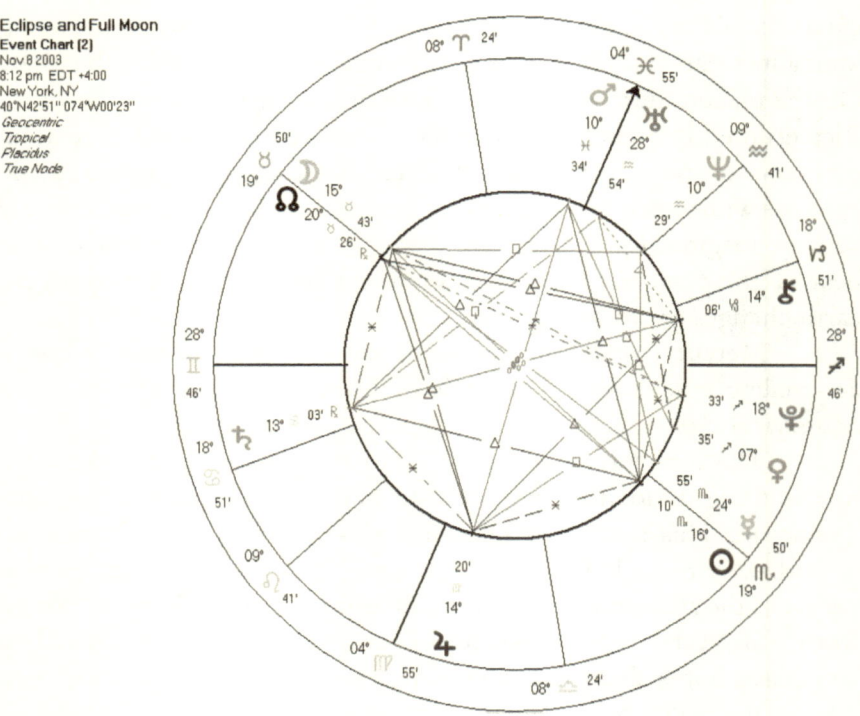

Figure 14. Eclipse and Full Moon Chart of Nov. 8, 2003

This configuration is a rare moment of beauty within the cosmic clockworks, unique not only because it is synchronous with an eclipse, but also because of the distinctive pattern of the outer frame—the two inverted triangles form the Star of David. This star has many symbolic meanings; the obvious one today refers to the State of Israel. The six-pointed star has been called the Seal of Solomon, but uncertainties surround that interpretation as some sources feel that this seal was a pentagram.

This star has been associated with sacred geometry as far back as ancient Egypt. Pythagoras was an Egyptian initiate who devised an entire metaphysical system using numbers and forms. Others trace it back to the Hindus. Some see it as a tantric symbol of male and female. For practical purposes, a useful interpretation includes a synthesis of the statement "As above, so below." The triangle with the top pointed downward suggests spirit, while the upright triangle signifies matter. Each mirrors the other.

In astrological terms, the pattern formed is a Grand Sextile. It signals an ease of flow between six planets through a sequence of 60 degrees. To be sure, there are squares, 90 degree angles, which typically signal stress, and oppositions, even more tension, in this chart. There are, to be exact, three oppositions and a T-square involving the Sun, Moon, and Neptune. The square between Venus and Mars ties in with the male/female, spirit/ matter theme and helps to set the stage for the transit of Venus the following June. A balancing of energies is suggested.

Moreover, the November full moon chart signals that there are stresses, but the Grand Sextile suggests a way out and through. This planetary configuration with its Grand Sextile in Water and Earth is distinctive for its rarity, encompassing as it does the elements of Water and Earth.

The full moon reflects the light of the Sun. At the time of the eclipse, we on Earth stand between the Sun and the Moon. This is a magnificent opportunity to reclaim our power of choice. The stars can only highlight areas where choice is most effective, either for the negative or the positive, depending on how we receive the energy.

This chart has to be tied in with the previous eclipse of May 15, 2003, which was also the Wesak Festival. This is the day, the full moon in Taurus, which every year celebrates the birth of the Buddha, the day that the Buddha attained enlightenment, and the day of his death. This November 8, 2003, eclipse is connected to the May, 2004, Wesak because it also falls at 15 degrees of the fixed signs—14 Taurus-Scorpio 42. And a month later, after the Wesak festival in the sign of Taurus, Venus transits the Sun.

This is one of those times when humanity can raise its consciousness and achieve enormous spiritual growth. There is much interesting information in the stars, and it would appear that humanity has and will continue to experience a wallop of energy relative to the 2004 transit.

Chapter Six

THE BEAUTY OF VENUS

We are not attracted to Venus because of mere superficial beauty. She is beautiful because within her there are certain prescribed qualities that naturally draw us. The Golden Mean is the key to a fuller understanding of beauty. It is essentially a ratio of approximately 1.618, to which the Greeks assigned the letter Phi. The Golden Mean is a mysterious, magical number which seems to be embedded in the basic structure of our universe. Phi appears in the realm of nature, in the realm of those things which unfold in gradual steps—this is the relationship that manifests in organic forms, be it a seashell or a pinecone. The actual decimal representation of Phi is 1.6180339887499 . . .

The 'Fibonacci series' gives an idea of the enticing character of these numbers. If you start with the numbers 0 and 1 and continue to make a list in which each new number is the sum of the previous two, you get a fascinating display of the relationship.

0, 1, 2, 3, 5, 8, 13, 21, 34, 55, 89, 144 . . . and on to infinity. If you take the ratio of any two sequential numbers (excluding 1/0) you'll find a range that oscillates around 1.6180889887499.

Even closer to home, the Golden Mean relates to the intricate unfolding process within the DNA of our own souls as we journey to higher states of consciousness. We are drawn to what is beautiful, and what is beautiful to us has an invisible connection to an implicit architectural design.

Psyche, as a co-relate to Venus, strives to attain what she desires. Psyche upon encountering obstacles strives to break out. She represents a whole stellium of psychic factors but is often equated with the Soul itself. Psyche is the inner core. As inspiration pours in, the core of the person becomes more open to new ideas and new visions. Such a person sees beyond the personality and appreciates the soul of others, even if they are hostile. In this way, Psyche learns to adapt, to understand. This is done by bringing in higher, and more refined frequencies with the resulting fusion to ever greater sources of cooperation and creative expression. The potential is limitless.

The Venus factor works magnetically—like the flower, she attracts the bee. Venus is magnetic, and draws to her. It is interesting to note that it has been said that wheat and bees came to us by way of Sanat Kumara, the Planetary Logos from the planet Venus. Honey is made from the nectar of flowers. Bees offer to man incredible life-enhancing gifts: raw honey, royal jelly, bee pollen, honey comb technology. Beeswax candles link us with fire. Moreover, it has been said that honey in its raw state has all the nutrients needed to be a complete food for human consumption. What a great gift we have from the bee, and thus from Venus.

Venus is associated with the 3rd trump card in the Tarot, as the Empress. This card stands for all things fruitful, giving birth, nature and the Feminine and most of all, Love. Tarot decks often show bees on the dress of the Empress linking her to Venus.

The process of creation and the process of formation speak to a resulting containment of something. The way in which the outcome is structured follows certain rules and boundaries.

It could be said that from the "radiant darkness" came all things in the universe. As the potential of the "radiant darkness" took form, there seemed to be some essential schematic into which all things fell. This underlying vibratory formula impacted all things in the universe— patterns, not only of molecules, but also of planetary and zodiacal energies. The formula for this pattern is found in the Golden Mean.

In the well-known drawing by Leonardo Da Vinci we get an idea of the symmetry inherent in the Golden Mean respective to the human body. From this drawing we can glean a sense of the modular relation of the parts to the whole. Socrates believed that it was the duty of the artist to present an idealized standard of the human body. Leonardo set out to do just that.

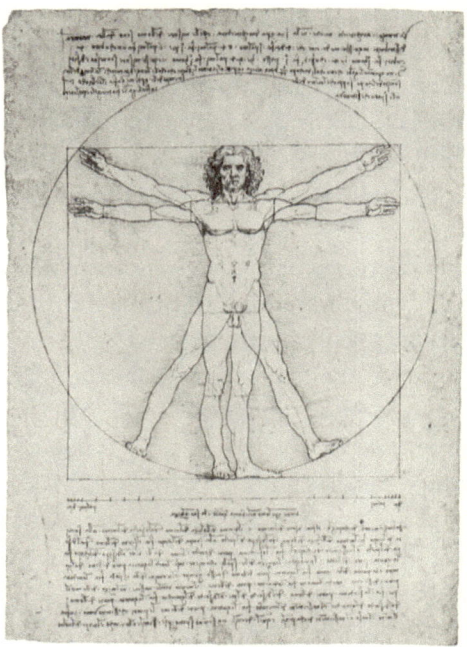

Figure 15. Symmetry of Man

Even in the animal kingdom, preference is given to those who present themselves in a beautiful symmetrical manner—the best plume, mane, or horns are noted. Symmetry matters. Form is designed in such a way as to have the potential to naturally attract. Proportion, design, color, all play a role in attracting our attention.

This fact takes a slightly different turn in forms from the ancient past. Venus figurines can be traced back as far as 20,000—27,000 years ago. Made of stone, wood, or ivory, a myriad of figurines have been found throughout Europe, from France to Russia. The most famous of these carved images is the Venus of Willendorf, so named for she was discovered in Austria.

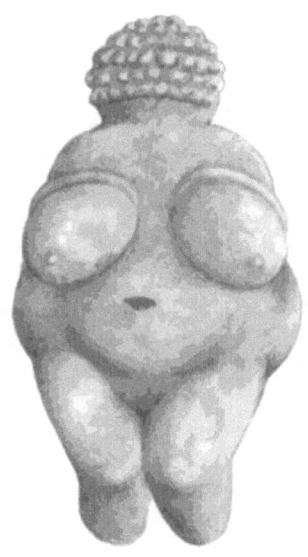

Figure 16. Venus of Willendorf

One glance at this beauty, with her swollen breasts and belly, tells us that she is pregnant. It is believed that this little Venus figure, along with the many others that have been found, were actually fertility objects or symbols of the Earth goddess. The tendency to create similar figures has been traced to even older civilizations to include Sumer. It has been suggested that these figures were some sort of universal symbol of an archetypal pattern which has been incorporated in the lives of Paleolithic people—one that speaks to deep, instinctual urges in the collective unconsciousness of humanity

The idea and image of Venus/Ishtar repeats over and over in ancient cultures. There is a heady acceptance of sex, fertility and the female form as a token of all things good. She represents fertility, abundance, sexuality. The cult of Ishtar was preeminent in ancient Babylonia as witnessed in the magnificent great Ishtar gates. The story of Ishtar's descent into the underworld in search of her husband was found on a clay tablet dating back to 1750 BCE. The inscription noted below gives insight to the personality of the goddess.

> My Lady abandoned heaven and earth to descend
> to the underworld.
> Inanna abandoned heaven and earth to descend to
> the underworld.
> She abandoned her office of holy priestess to
> descend to the underworld . . . .
> If I do not return,
> Set up a lament for me by the ruins.
> Beat the drum for me in the assembly places.
> Circle the houses of the gods.
> Tear at your eyes, at your mouth, at your thighs . . . .
> Clay tablet

In this rendering Ishtar/Venus is called Inanna the Sumerian goddess of not only love, sex and fertility but also of warfare. This association may explain why negative predictions were associated with Ishtar in the ancient past.

What is important is the awareness of an innate tendency on the part of humans to express and to create. It would appear that this impulse to fabricate and relate to these delightful figures, which represent fertility, fecundity, and the feminine is deeply and undeniably embedded in human consciousness. The ancient past embedded and expressed in their form is still provocative. There is, undeniably, an attraction as well as a curiosity to them even today.

The figures may represent early people's attempt to personify their creator as mother—a tie to what had developed as matriarchal societies. These early precursors of a more sophisticated and developed Venus share characteristics with the Gaia figure in ancient Greece, and with the later Roman and Greek classical Venus and Aphrodite beliefs. They share common themes—sexuality, love, fertility, and nudity. Rarely is a Venus clad. Down through the ages her form is laid bare for all to see and appreciate. Sensuality, but not necessarily sexuality, is equated with Venus.

Venus, and its association with peace, plays an important role in the unfoldment of the soul. The eternal beauty that we seek and long for is not of this world. Even in the Venus of Willendorf there is the sense of the primal and the eternal, the light of life itself. This is the light that Venus carried and continues to carry—the extension of the light of the divine. With it comes a sense of mystery and of things that we will never completely fathom.

> Beauty is unbearable, drives us to despair,
> offering us for a minute the glimpse of an eternity
> that we should like to stretch out over the whole
> of time.
>
> *Notebook,* Albert Camus.

It is this glimpse into the eternal that Venus offers. There are physical correspondences. She is depicted naked because she reveals secret things. With her radiant beauty, untroubled youth, and tender charms, her every step spreads the scent of ambrosia to the delight of her surroundings. It is said that a delicious scent filled the air around the goddess. With all her charms and tempting attributes, Venus personifies one thousand variations of love. She has the power to awaken slumbering desire—even the purest of heart would succumb.

The girdle of Venus is said to awaken the senses. Even the great goddess Juno had to borrow it when she wanted to gain approval from the Lord of the World when she found herself with him on Mt. Ida. Even Jupiter couldn't gaze at Venus without being moved. Venus ruled over the all the Great of the World.

Pliny wrote:

> Everything on earth is generated by her power.
> Through her twofold appearance—
> Once in the morning and again in the evening—
> She scatters fertile dew on the earth and animates
> all living creatures and multiplies.

In the metallic kingdom, we associate copper with Venus. Copper wires are used to support electric currents. But, of course, there are no copper wires in outer space. James Maxwell, a Scottish scientist, proposed that electricity could travel through space, but only with a very fast oscillation. He believed that with a displacement current added to the equations of electricity and magnetism, one could propagate something equivalent to the speed of light, and light was eventually seen as an electromagnetic wave. Similarly, the attractive principle of Love corresponds to magnetism.

Assuming that the first creation out of the "radiant darkness" was Light, the Creator or Cosmic Logos had the impulse to individualize units of light—what we call I Am. Presumably the "I Am" factor exists beyond time and space. Einstein reached and never completely succeeded in defining this area in his General Theory of Relativity, and even later when he attempted the Theory of Everything, he stumbled into and encountered mystery.

> *The most beautiful experience we can have is the mysterious the fundamental emotion which stands at the cradle of true art and science.* —*Albert Einstein.*

Einstein clearly made the link between beauty and science. Perhaps beauty is really about the presence of magic in our lives. True, science and the spiritual often collide in the kind of questions that naturally arise. Like beauty, the art of science also appears magical as it works itself out in our lives. Within this milieu, science and metaphysics often resort to the language of symbols in their mutual attempt to uncover the meaning of primal imagery within the interconnectedness of all creation.

In the painting of the Marriage of Eros and Psyche, there are the qualities of innocence with beauty that stand out. Venus stands by, and hides the bow that Eros used so effectively on others. The painting is not sophisticated or mechanistic. It provides a level of experience that is extreme, heightened, and awake. Love opens the heart, and with that opening comes an inflow of energy. Love allows us to see the beauty in nature and in each other.

With its aura of mystery and magic, and its association with the beautiful, the fairy tale of **The Sleeping Beauty** is illuminating. It suggests that there is an innate impulse in all people to raise their own consciousness, break through obstacles, and embrace the beauty that sleeps, not just in one beautiful human being, but in all humanity. Poor Beauty—she has been imprisoned for 1000 years, bound to a consciousness that held her dormant and in a state of inertia. The prince is that part of ourselves that must hack and bleed his way through the briar and brambles, almost not making it, but willing to encounter difficulties in order to finally arrive where beauty sleeps. With a kiss he wakes her up.

The tale shows how we unfold in the direction of the beautiful *from death to immortality, from chaos to beauty*. Peace and conflict reside within. The 2004 transit suggests experiences that can shed light on separatism and warring thought systems. War is a possibility, but not a necessity if the struggle works out on the mental plane. Moving, as we are, toward a system of cooperation, peace, and harmony will require a release of beauty. Beauty has been asleep. Once awake, her divine light can and will catalyze enormous changes. The job of humanity, at this time, is to save ourselves from sleep. The prince is that part of us that is compelled to strive. He will not hesitate to use magic to create the awakening.

A desire to release beauty within ourselves and within others can be energized with the Venus transit. Bonds of materialism, separatism, ignorance, and inertia, once dissolved, will allow humanity to move on to a fresh connection with the light and engender a continuity of consciousness. The magnitude of beauty will increase as cosmic fire progresses though our system. Magnetism and expression of beauty will emerge and create deep changes in human awareness.

The secret of beauty is not necessarily about form—it is about the structuring and restructuring of energy. We see a beautiful object, respond, and something shifts in us. We resonate with something higher, more refined and more subtle.

Chapter Seven

VENUS AND THE FIFTH RAY

The 18th century artist Boucher's delicate, lighthearted depiction of a gorgeous nude Venus is one of sweet natural beauty surrounded by luscious colors and textures representing brocade, satin, and silk—fabrics with which Venus can, if she chooses, veil herself. But the goddess is relaxed, certain of her gifts, and without a trace of shame or regret and she remains, certain of her beauty and innocence, unclad. The white dove of peace and purity nestles at her feet. Little angels are her companions—one toys with her blue ribbon, another arranges her hair, and still another plays with a long strand of luminous pearls. She is beauty infused in form. Beauty and humanity can know each other; through the beauty of Nature's mysteries, we can know of the Divine.

From Boucher's Venus we can glean a sense of the weaving of the Immortals. Patterns are shown; they not only can be treasured but studied. From this experience, the secret of how truth can be approached is revealed. As we continue on the path of evolution or growth into light, there comes an awareness of the changeless beauty of all things. A creative life or a creative experience yields a type of participation in the continuation of the process of Divine creation. A truly creative person or a creative act allows the Divine to show forth it power and beauty. Themes of beauty, art, creativity and higher mind relate to Venus and her relationship to earth and to humanity.

The very beauty of the solar system can be viewed as a Divine garment, a web of Beauty. Just as in Boucher's painting, truth lies hidden in the folds, strands, and ribbons of light within the garment of the universe—galaxies, constellations, and solar systems. The painting offers insights into the process of divine manifestation.

Figure 17. Boucher, Toilette of Venus

Art conditions the mind not only to expand but to refine. Boucher's painting suggests new interpretations of Venus and also re-enforces many of the old. Through color and design the artist leads. It has been said that the fourth, the Atlantean race developed the emotional nature and the expression of beauty and color through art. We carry over what was gleaned from the past. Now, it is our turn, the task of the fifth root race, to build on that legacy. Having incorporated the lessons from Atlantis, we can subsequently bring in the creative power of the mind and with it, the ability to analyze, to scrutinize.

What possible relevance has this to do with our current world? Venus is associated with Ray Five, the Ray of Science and Technology. It is our task to elevate the creative power of the mind in the realm of science and technology. Linked to a higher purpose, a purpose that transcends material benefit, the light of the mind can be raised to a point of maximum brilliance in the service of humanity.

Esotericists describe a system like ours as a triple solar system. The Christian perspective is similar—three Persons in One God. We can say that a Divine mantle of energy manifests as a three-fold system, as rays or streams of energy. These streams have definite characteristics and purposes, and are the seven great builders of all that exists.

> We may ask, what is a Ray? A Ray is but a name for a particular force or type of energy, with the emphasis upon the quality which that force exhibits and not upon the form aspect that it creates.
>
> *Seven Rays Made Visible.*
> Helen S Burmester. P.9.

The First Ray represents Father, life, and will. The energy is positive.

The Second Ray represents Son, Consciousness, Love-Wisdom, and balanced energy.

The Third Ray represents Holy Spirit, form, Active Intelligence, and negative energy.

The Absolute Differentiates into the Three

These three rays are present to some degree in all things. In a way, they are analogous to Shiva, Vishnu, and Brahma and/or the three gunas in Hindu philosophy. They can be taken to mean the three aspects of God, the three aspects of all manifestation. As the energy from these three moves and circulates, it is further expressed into seven rays in all—three major and four minor. The minor rays issue forth from the Third Ray. In total, then, there are seven rays. There are seven centers, sources of a specific kind of energy, or chakras in the human being and in cosmic systems. The characteristics of the seven source-forces may be described as follows:

Ray I	Ray of Will or Power, 1st Aspect
Ray II	Ray of Love-Wisdom, 2nd Aspect
Ray III	Ray of Active Intelligence, 3rd Aspect

(from the above three come the following;)

Ray IV	Ray of Harmony, Beauty and Art
Ray V	Ray of Concrete Knowledge or Science
Ray VI	Ray of Devotion and Abstract Idealism
Ray VII	Ray of Ceremonial Magic or Order

Table 2. The Seven Rays

> Divisions and categories are *not* separations, because all life is one—one organism, in which every cell plays its part. But these divisions help us to understand more clearly and intelligently the issues involved.
> *The Seven Rays Made Visual* Burmester p.9

We are part of a Greater Whole and within that system the entire phenomenon of the vault of the heavens still poses many mysteries. To better understand these mysteries, humanity has studied the stars and tried to understand how the many diversified forms function and relate to each other. Scientists like Newton, Kepler, and Galileo have shown us that a law of periodicity functions in the heavens. It has been used to calculate and reveal the nature of the orbits and cycles of the heavenly bodies. Thanks to Galileo's efforts we can successfully predict the transits of Venus. But even before Galileo, the Mayans made incredibly accurate observations, using complicated mathematical cycles of time, they could project ideas and plan for the future. They used shards of black obsidian to protect the eye as they looked directly into the Sun, learned about Venus and made notations.

In the esoteric scheme of things there is a similar law. It is called the Law of Periodicity. It relates to all things in manifestation—solar systems, universes, and the kingdoms of nature. Within cyclic-rhythms, creativity advances. A knowledge of these cycles can be helpful, particularly to groups as they engage in building new and better cultures and civilizations. Those people who are so engaged and are in the process of responding to refined energy are known as the New Group of World Servers. They function within a three year cycle. Their cycles begin with the Full Moon in Taurus, with the great downpour of energy that is released for humanity at that time.

First Year	Second Year	Third Year
May 1990-1991	May 1991-1992	May 1992-1993
1993-1994	1994-1995	1995-1996
1996-1997	1997-1998	1998-1999
1999-2000	2000-2001	2001-2002
2002-2003	2003-2004	2004-2005
2005-2006	2006-2007	2007-2008
2008-2009	2009-2010	2010-2011

Table 3. Cycles of Group Work

The rhythmic flow of this creative cycle allows the New Group of World Servers, whoever they may be, to plan ahead and to make themselves more efficient. They are likely to make fewer errors and to go with the flow of energy in a productive manner.

Looking at the table above we can see three waves or impulses of energy. They can be clustered in the following manner.

- Consolidation: First Year: Activity related to the gathering in of the needed ingredients for manifestation of whatever the task will be.
- Expansion: Second Year: Refine and fine-tune the process.
- Outpouring: Third year: Pouring out life and energy into the work.

In addition to the three year cycles for group work, there are larger cycles. A complete and fascinating discussion can be found in *Discipleship in the New Age*, Volume I and *Esoteric Psychology*, Volume II, both written by Alice Bailey. In terms of the larger, and more expansive Transits of Venus, these shorter cycles indicate steps or methods along the way that can be employed to bring about desired effects.

In brief, the larger, seven year cycles correspond to the particular energies coming to our earth by means of the constellation Capricorn. To be specific, these energies, which have been said to have begun flowing in 1935, will continue to flow every year during the week of December 21. If a full moon should happen to coincide within that week, the opportunity would only be enhanced. These unique opportunities have been designated as the "festival week." It is hoped that by providing this information, groups find this knowledge useful, and if they do, they can take advantage of the information that has been given regarding the cycles. The concept of group work is gaining momentum as we move into the Age of Aquarius. It is through groups that the waters of life, which Aquarius brings, will be distributed on Earth. The following Table provides information to guide groups in planning future activity.

December 21-28
1991—1998
2005—2012
2019—2026

Table 4. Seven Year Cycles of Group Opportunity

The 2012 Cycle of Group Opportunity precedes the 2012 transit of Venus by less than six months. This particular December 2012 festival should provide great impetus for group work.

A single chart has been chosen to exemplify the individuals who clearly belong to and have participated in the New Group of World Servers. They are pathfinders on the way. The chart of Martin Luther King provides a glimpse into the dynamic life of one individual who was willing to risk all to further the plan—one who responded to a fiery call with an equally fiery will. Even if the reader is not interested or versed in astrology, we invite the reading of the explication for it carries subtle messages in words and symbols. Here is a depiction of a vibrant, living experience of one individual who took a stand for light.

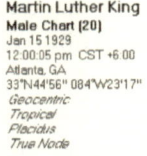

Martin Luther King
Male Chart (20)
Jan 15 1929
12:00:05 pm CST +6:00
Atlanta, GA
33°N44'56" 084°W23'17"
Geocentric
Tropical
Placidus
True Node

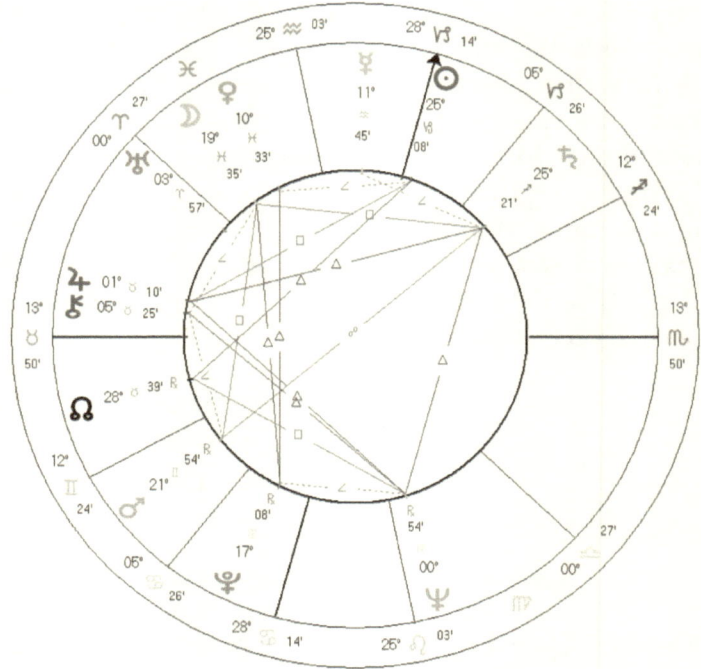

Figure 18. Chart of Martin Luther King

What immediately stands out in this chart is the Capricorn Sun, the sign of an initiate, at the Midheaven. From this lofty position, the Sun radiates through the entire chart. Fire is the means of development, and in this case it illuminates the entire life pattern.

The Sabian symbol for 26 degrees Capricorn is: "A Nature Spirit Dancing in the Iridescent Mist of a Waterfall. The Keynote: *The ability to perceive the hidden and creative spirit of natural phenomena.*" Rudyhar, *An Astrological Mandala*, p. 245.

Here we are dealing with revelation. King's mental capacities and his skill in communicating reveal a fluidity of consciousness, and in this case a fluidity which finds itself stimulated by conflict. The soul of nature bears a correspondence to the Soul of the World. King was not only sensitive to the downward flow of energy which had been made available to him, but he was able to work with those energies to make magical transformations on a global scale.

Our solar system is a chakra within a larger body, and within that body the Sun is the heart charka, the "jewel in the lotus," the inner central core of electrical fire and light. The jewel has been likened to the Eye of Spirit. In this chart the Eye of Spirit observed and maintained contact, peering down into the fiery life of this man who became a martyr to truth.

Djwhal Khul, the Tibetan Master speaks of fire in this way.

> **Fire:** Fire internal and latent; fire radiatory and emanative; fire generated assimilated and radiated; fire vivifying, stimulating and destroying; fire transmitted, reflected and absorbed; fire, the basis of all life; fire the essence of all existence; fire the means of development, and the impulse behind all evolutionary process; fire the builder; the preserver and the constructor; fire, the originator, the process and the goal; fire the purifier and the consumer. The God of Fire and the fire of God interacting upon each other, till all fires blend and blaze and till all that exists, is passed through the fire—from a solar system to an ant—and emerges as a triple perfection. Fire then passes out from the ring-pass-not as perfected essence, whether essence, emerging from the human ring-pass-not, the planetary ring-pass-not or the solar. The wheel of fire turns and all within that wheel is subjected to the threefold flame, and eventually stands perfected. Bailey *Treatise on Cosmic Fire*, p 97

Within the context of this chart we can see the importance of Solar Fire. With Taurus on the Ascendant, King experienced a life that permitted the focused beam of light from Taurus to stream forth, through the Eye of the Bull and reveal the path. As stated earlier, in esoteric astrology the Ascendant describes the Soul's purpose. Venus, the ruler of Taurus, makes a trine to Pluto. Yes, Venus and Pluto make things happen, and in this case Martin Luther King exemplifies one who was able to walk into darkness to irradiate and uplift the physical plane. This chart presents a good example of desire transmuted and Venus activated. The resulting spiritual aspiration (desire transmuted) made an impact on the entire world. Through the Eye of the Bull appears the path of detachment from desire, and with that detachment comes the awakening of the third eye. As above, so below—the Eye of Spirit looks down from above and in response, the awakened individual brings forth the fullest expression of Soul. The Creative Will comes into play, bestowing strength and persistence.

Mercury in the Tenth House is also prominent in the chart. With Mercury in Aquarius, Brotherhood is expressed in daily life and in a public forum. Mercury also cooperates with the theme of light for it represents intuition. The kind of awakening that Taurus brings reflects a consciousness that is Soul centered and draws upon intuitive knowing—straight knowledge.

Taurus is found on the Fourth Ray of Harmony through Conflict. The Fourth Ray corresponds to the Fourth Chakra—the heart charka. It is through the heart that the message of Love, the message of Christ emanates. Anyone who has heard the "I Have a Dream" speech will recognize Martin Luther King as the Christian minister who overflowed with the message of love and the quality of fierce Will. The heart is Love, the brow is Wisdom, and the head is Will—an integrated soul-infused person whose design fits into the bigger scheme of things. It must be remembered that Taurus very much relates to Earthly life. The Fourth Ray of Harmony through Conflict is the ray of humanity; this is not only a life of great conflict but also one of enormous service to humanity. To master conflict, one must become adept in the art of living, because the trials and tests must be encountered and passed. The mettle must be tested.

The Fourth Ray is also associated with Art and Beauty, so this is a life of great Beauty. The poem of William Blake comes to mind when pondering the life of Martin Luther King.

> *Tiger, Tiger, Burning Bright*
> *In the Fires of the Night*
> *What Immortal Hand or Eye*
> *Framed thy fearful Symmetry.*

Here is a horoscope with fearful symmetry. That it is congruent with the plan is obvious. The design is of great Beauty, but it is terrifying because Beauty demanded all.

One more point about the dynamics of this life. It must be remembered that Venus brings the Fifth Ray or lower mind to the picture. Venus gave the gift of mind to animal man.

> Man is a living entity, a conscious son of God (a soul) occupying an animal body. Here lies the point. He is therefore in the nature of a link and a far from missing link. He unifies in himself the results of the evolutionary process as it has been carried on during the past ages, and he brings into contact with that evolutionary result a new factor, that of an individual, self-sustaining and self knowing aspect. *Serving Humanity*, Bailey, p.53.

It is the gift of mind, of awareness, that was the gift of Venus. It is this very Venusian energy which stimulates the Taurus type toward greater ascendancy through the agency of the intuition, which eventually happens when the third eye finally opens. Martin Luther King exemplifies the disciple who was able to transmute earthly desire into Spiritual Will and become a true initiate. There is movement in the chart and in the life. The third eye opened and a unification took place.

Both Jupiter and Chiron trine Neptune—a planet that cumulatively floods the chart with the Light of Love. Jupiter, the ruler of the Eighth House of Death, is conjunct Chiron. And so it was that this great spirit was wounded, no doubt with karmic ramifications. This is indicated because both Jupiter and Chiron are in the Twelfth House, the House of Pisces. Both the past and what may manifest in the near present can be seen. Sacrifice is certainly implied.

Vulcan cannot be ignored in a chart with a Taurus Ascendant, since Vulcan is the partner of Venus. We can locate Vulcan within 8 degrees 30 minutes of the Sun's position on either side of the Sun. In the case of Martin Luther King, it is safe to presume that Vulcan is very close to the Midheaven. Vulcan carries the First Ray, as does Pluto. We have already commented on the Venus/Pluto trine and now we are considering the ramifications of Midheaven/Vulcan. The positive effect of this tiny planet is expressed in this chart as the Will-to-Good. Solar Fire is radiating creative potency with the Sun and Vulcan providing the force of Divine Will.

Of Vulcan we know this—Taurus forges the instruments of constructive living, or of destruction. Vulcan forges the chains which bind, but it can also create the key which unlocks the mystery of life. Vulcan has earned the reputation of the "Blacksmith of the Gods."

Vulcan provides the wherewithal to create the tools and weapons to become a builder in the light. Vulcan proves the metal which will be tested and tried with various challenges, all of which revolve around detachment. These themes bear directly on the chart and the life of Martin Luther King. He had a premonition that he would die, and he accepted that challenge willingly. The tools he used were Christ consciousness, loving understanding, dynamic powers of speech and action, and the ultimate expression of Love, to forfeit his life for humanity.

King's life and pattern present a good opportunity to study the mechanics of esoteric astrology. It also shows what Venus and the Venusian influence can accomplish. Taurus qualities of Right Speech and Right Livelihood are attributes of the true white magician, the true teacher, and the true healer—released, transcendental, and group aware.

King's life stands out as an example of the kind of bridging work that needs to be done. The transits of Venus, which is the over-riding theme of this book, can also be seen as bridging opportunities—bridging between where we stand today and the kind of future we wish to create. Within the big picture, King, with this Taurus influence, stands as a harbinger of things to come.

One of the mantrams that has been employed by group workers follows:

The Great Invocation

From the point of Light with the Mind of God
Let light stream forth into the minds of men.
Let Light descend on Earth.

From the point of Love within the Heart of God
Let love stream forth into the hearts of men.
May Christ return to Earth.

From the center where the Will of God is known
Let purpose guide the little wills of men,
The purpose which the Masters know and serve.

From the center which we call the race of men

Let the Plan of Love and Light work out.
And may it seal the door where evil dwells.

Let Light and Love and Power restore the Plan on Earth.

The Great Invocation allows individuals or groups a way to invoke those energies which can be directly and positively used to make needed changes on a global stage. Love, Light and Power are consciously invoked and an evocative response can issue forth. Often used at the time of the Full Moon celebrations, a time when certain planetary energies are particularly available, groups can say the Great Invocation then direct to problematic areas in the world.

It is useful that astrologers, as well as all interested in the Path of Ascent, pay significant attention to the methods by which thought can be refined and Soul contact made real. With this in mind, it is imperative that the refinement of thought must be developed along with technology as civilizations and cultures change and evolve. Here is yet another reason to better understand the Venus transits and the advantages that will come with it.

Down through the ages, there have surfaced The Seven Laws of the Soul—they underlie all spiritual and religious teachings. These are certain basic underlying truths laid out in Esoteric Psychology, Vol. II by Alice Bailey. We know, for example that it is the stuff of matter that the soul has to employ and that matter can and will be reinforced, elevated and transformed. Karma, daily life, and entropy press against the soul's purpose and desire to evolve. But with the right use of time, understanding, and love the underlying inertia can be overcome. The Seven Laws of the Soul provide not only a foundation but suggest an ultimate destination for Soul progress—a state of fiery liberation. The laws provide a matrix for growth and progress as well as to highlight the great alchemical process of White Magic—the process that spiritualizes matter and all forms of substance.

The Laws are stated here to stimulate interest and to suggest a relationship to the Seven Rays. Each law of the soul correlates to one of the major rays. A more detailed study will reveal that the chakras can and must be included in the overall strategy of soul development. Chakras, those seven primary centers found along the spin, are the chief energy centers through which forces move and stimulate evolution.

A knowledge of these laws will enhance their usage and application to the world at large. They are listed below.

The Law of Economy—governs matter and the third aspect
The Law of Attraction—governs the soul, the second aspect
The Law of Synthesis—governs spirit, the first aspect.
The Law of Vibration
The Law of Cohesion
The Law of Disintegration
The Law of Magnetic Control
The Law of Fixation
The Law of Love
The Law of Sacrifice and Death

Table 5. The Laws of the Soul

The seven laws of the Soul use time with focus and with skill. They interrelate to the Seven Rays within any given period of manifestation, each Ray sweeps periodically into power and then subsides into a period of quiescence. The mystery of the cycles had been of enormous interest to scientists, metaphysicians, and artists since time began. Complete knowledge of the great mysteries has been the goal of seekers and white magicians, and has been the attainment of the adepts. White magicians are those who understand and consciously work with the soul in order to bring forth soul infused energy. This involves the manipulation of energy and matter in such a way to make it more refined, and more useful for evolution in general. White magic deals with both the microcosm as well as the macrocosm. Its goal is both the spiritual and material evolution of humanity.

Since the Middle Ages, the Catholic Church had devised a grand system of correspondences. God and the Divine were seen as the macrocosm within which we function in our microcosm. Man was depicted within this set-up as superior to the plants and animals, which had been put in his care. Man, of course, owed his allegiance to God, and an elaborate hierarchy was created—the great chain of being.

Figure 19. Rhetorica Christiana

The powerful image of a divinely inspired hierarchy provided a structure on which all forms of life of life could be ranked. In this rendering by Didacus Valades, from *Rhetorica Christiana*, 1579, only the male is depicted. The representation was reproduced in Anthony Fletcher's *Gender, Sex and Subordination*.

It is interesting to note the emphasis on the male in this hierarchical set-up. Man, and man alone, is represented in the fire of life. What happened to Eve? She is notably missing in this rendering. The 17th century transit of Venus changed all that. The influx of artistic renderings exploded. They depicted not only the feminine, but Venus, naked in all her glory.

The medievalist and the metaphysician would agree that mankind is certainly an expression of God. The addition of the system of seven rays provides a schematic within which one can be more exact in terms of discussion and depiction. Humanity progresses through self-realization and self-expression using the energies of the seven rays. Where, then, does Venus stand in this structure of things?

"Venus has been associated with Ray Five. Venus stands for the emergence of the love principle through the directing power of the mind." *Esoteric Astrology*, Alice Bailey, pp126-7.

Through the energies of Venus, the soul and the personality come under the sway of the mind. Eventually, mental energy can be transmuted into wisdom. As seen in the Sleeping Beauty legend, the release factor lies in the hands of love. The task of Venus is to unite the personality and the soul, producing the ultimate unification of heart and mind.

Venus is the illuminator, the source of intelligent mind. The Fifth Ray, associated with concrete knowledge and science through Venus, brings the human being to a point where growth is possible. In a way, she is considered Earth's alter ego; she is so like us and yet, at the same time, complimentary. Older than her sister Earth, she is our primary prototype. With her age and experience, she can assist the Earth to release sorrow and pain, and thus reveal the inner beauty.

Today, esotericism—a complementary stream of thought which uses an elaborate, hierarchical system to understand the structure of life and its purpose—demonstrates the spiritual side of Venus. She, with her mate, Vulcan, typifies the form and the soul. It is understood that Venus is the planet which was responsible for the appearance of the individualized consciousness in man. No small feat. Bailey asserts that Venus (along with Pluto) are dominate in bringing about world results.

There is this underlying truth—there is one life, and it expresses itself through seven basic rays, resulting in the myriad of forms that we know. The Rays can be taken to be lives, entities who infuse form with meaning and encourage the impulse to know and grow. In a chart, the astrologer looks to the Second House for clues. The Second House is naturally the house of Taurus and is strongly influenced by Venus— Taurus the mother of illumination and Venus the giver of mind plus the embodied soul, as well as the distribution of money and metals. Venus is one of the planets that furnishes us with our outward and inward characteristics. She is associated with the Holy Spirit, one of the Four Guardian angels of the Earth, and can be considered the Light Bearer on Earth.

Ray V—Concrete Science of knowledge and technology—comes to us via Venus.

The country of France has a Fifth Ray soul. As that country cooperates more and more with its soul ray, the Fifth Ray, astonishing accomplishments can come about.

"Through the French nation—a consummation of the Piscean influence or genius." *The Destiny of the Nations*, pp 74-74.

Perhaps this transit will bring to France the last needed bit of energy to lift its consciousness and truly fulfill a destiny that is still yearning to be complete. It has been suggested that it will be the French who will discover scientific proof of the soul. They, collectively have the ability to do so. Moreover, the French role and commitment to a unified Europe is key to European success—diversity within unity.

Each ray life is uniquely and distinctively expressing itself. The seven rays flow through each of the seven planets and predominantly through one of what we call the sacred planets. These are planets that have achieved a significant level of development in the scheme of things.

The Sacred Planets

Vulcan	Ray I
Jupiter	Ray II
Saturn	Ray III
Mercury	Ray IV
Venus	Ray V
Neptune	Ray VI
Uranus Ray	VII

Non-Sacred Planets

Pluto	Ray I
Sun	Ray II
Earth	Ray III
Moon	Ray IV
Mars	Ray VI

Table 6. The Sacred Planets

Esoteric astrology designates both the Sun and the Moon as non-sacred planets. It is said that they reflect light from certain sacred planets which they veil—the Sun veils Neptune and Uranus, while the Moon veils Neptune.

In a horoscope, each sign has a designated exoteric and esoteric ruler. In brief, exoteric rulers influence the mundane aspects of life. Esoteric rulers influence spiritual development, and can indicate the opportunities available to more advanced individuals as they respond to those higher energies that become available along the way. Every month, as the Sun progresses through the signs, the personality as well as the soul of man responds to those energies which are made available.

Rulership can be seen in this manner:

Signs, Planets and Rulers			
Sign	Exoteric Ruler/ Rulers	Esoteric Rulers	Hierarchical Rulers
Capricorn	Saturn	Saturn	Venus
Aquarius	Uranus	Jupiter	Moon/Uranus
Pisces	Jupiter	Pluto	Pluto
Aries	Mars	Mercury	Uranus
Taurus	Venus	Vulcan	Vulcan
Gemini	Mercury	Venus	Earth
Cancer	Moon	Neptune	Neptune
Leo	Sun	Sun/Neptune	Sun/Uranus
Virgo	Mercury	Moon/Vulcan	Jupiter
Libra	Venus	Uranus	Saturn
Scorpio	Mars	Mars/Pluto	Mercury
Sagittarius	Jupiter	Earth	Mars

Table 7. Signs, Planets and Rulers

A single human, a group of people, or, on the larger scale, humanity considered collectively can progress to the degree that the personality and soul level of integration permits. The inner guide, the soul, reveals itself gradually. There is, of course, danger passing from the old, crystallized ways of thinking and being into the new. Revelation of this kind is related to transformation. Substance is actually changed. The higher the degree of transformation, the greater becomes the ability to receive and use higher energies until that day when one is freed from constraints and truly becomes "enlightened."

The spiritual Hierarchy refers to those enlightened ones who have advanced beyond humanity, but who offer radiation and guidance to those who are ready. Hierarchical rulers, in effect, are constantly sharing their light. It is only our limited capacity to actually incorporate them and employ them that prevents their fuller use. Still, we call them out, i.e. Hierarchical Rulers, in a chart so that individual can ponder new opportunities that might become available.

Our planet is not yet a sacred planet because certain conditions have not been met. Great teachers like the Christ, as well as many great ones from the East, have brought the message of perfection to humanity. Within the scheme of things, Mohammed played an important role.

It would appear that the emphasis laid by the followers of Mohammed upon the fact of God, the Supreme, the One and Only, was in the nature of a balancing pronouncement, coming forth as it did in the fifteenth century, in order to safeguard man from forgetfulness of God, as he drew nearer to this own latent and essential divinity as son of the Father. The study of the relationship of these different faiths, and the manner in which they prepare for and complement each other, is of the deepest interest. This our Western theologians have often forgotten. Christianity may and does preserve secret within itself the sacred teaching, but it inherited that teaching from the past. It may personalize itself through the instrumentality of the greatest of the divine Messengers, but the way of that Messenger had been prepared beforehand, and He had been preceded by other great Sons of God. His word may be the life-giving Word for our Western civilization, and may embody the salvation which had to be brought to us, but the East had its own teachers, and each of the past civilizations upon our planet had had its divine Representative. As we consider the message of Christianity and its unique contribution, let us not forget the past, for if we do we shall never understand our own faith. From Bethlehem to Calvary. Alice Bailey pp. 33-34

Over time, the influences of the interacting energies and influences coming through the great cosmic pattern will eventually transform our planet, the Earth, into a sacred planet. This is our destiny.

The uniquely refined and powerful energies from the sacred planets stimulate our own growth on Earth as we learn to integrate them into our own being and activity. In this manner we work out the process of evolution—at first on the individual level. Later, it is accelerated by group work, and eventually impacts the planet Earth as a whole. Every planet plays its perfect part in the overall plan of life. Every planet is congruent with the purpose and the intent of life.

Earth, as has been stated, is not yet a sacred planet. In fact it may be viewed as a battlefield where the two first aspects—will and love—are in collision. It is through the instrumentation of white magic that eventual triumph will occur.

At present Venus, Jupiter, and Saturn are the principle vehicles for the three principle Ray aspects. This is reason enough to pay attention to Venus activity—it is compelling and worth the effort. A Ray is simply the name of a particular force or a type of energy which manifests in a particular by that ray.

Venus connotes to our minds, even if we have only a glimmer of occult truth, that which is mental, that which concerns final sublimation, that which deals with sex and that which must work out into symbolic expression upon the physical plane. These are the major concepts which enter our minds when Venus and Taurus are considered in unison. *Esoteric Astrology*, p. 384.

Chapter Eight

Cycles Of Light—Past Transits

The purpose of this book is to offer an additional and evocative body of information to complement the scientific and technical data from the hard sciences. Along those lines, working under a time pressure in order to get the information out in a timely fashion and into the hands of readers in time for the 2012 transits, the historical survey offered here is only rudimentary, and focuses primarily on the Western world. This is especially evident in the regrettable lack of information regarding the East—events, issues, and circumstance—from which we could extrapolate possible, potential events. The first transit impacts the Eastern part of our globe significantly, and it is worth the time to consider those areas in detail. We solicit the input of readers to contribute online to our historical database so that the general bank of information can be more fully enhanced and our learning enriched.

We are interested in outstanding contributors to the world of thought, art, and historical trends, or events on the world stage. These can coincide with dates of birth and death, or the interim periods between transits in which notable events occur. Such happenings may have triggered far reaching trends or movements, as is the case with Susan B. Anthony and her notable stand for women's right to vote in 1872. This is an event which led to universal suffrage and many other related contributions of women for humanity.

Admittedly, this is a rudimentary approach, but at the very least, it may trigger interest in the significance of the transits and encourage further research. In the meantime, the following events have been noted.

December 1631-1639 Transits

1631:   First newspaper, Gazette of France, was published. It is still in circulation.

1631:   Mir Damad, Muhammad Baqir, died. Primarily an agnostic writer, he was referred to as the 'Third Master' after Aristotle and al-Farabi. Mir Damad was a jurist, a mystic, and a philosopher. He interests were comprehensive; his style was suggestive, symbolic, and referential, relying heavily on a thorough knowledge of existing Islamic philosophy. Like Plato, he was a realist, accepting the materiality of being. *Al-Qabasat* is mir Damad's most significant philosophical work.

1631:   John Donne died, but the great tradition of English metaphysical poetry reached a summit with Henry Vaughn, George Herbert, and Richard Lovelace.

  In Art there were Bernini, Franz Hols, and Rembrandt—masters whose created works have never been equaled in form, delicacy of detail as well as dramatic power. Colors deepened, moods were emphasized, and technique reached a new level of excellence.

1632:   Shortly after the first 17th century transit and during the interim between transits, two great contributors to the world of thought, Benedict Spinoza and John Locke were born.

  Spinoza was a major contributor to philosophy. He modeled his approach after geometry and addressed issues that concerned the nature and the origins of the human mind. In his pioneering work he analyzed and wrote concerning emotions as well as human bondage.

  John Locke, a major influence in English thought, was not only interested in philosophy but also mathematics. He studied and wrote on the origin of ideas. His concepts on the mechanics of human understanding, like Spinoza, were grounded in traditional philosophy. Locke explored the nature of the self and of God as the basis for knowledge, and came up with the unconventional idea that government is a social contract, a notion that would feed into and nourish the Age of Enlightenment.

1639:   John Milton, a major luminary of English literature matriculated at Christ College. In 1638 he published his now famous elegy,

Lycidas. In 1630, he moved to London where his letters and poems had enormous influence, not only artistically but also politically. In 1667, long after the transit of Venus, Milton, by that time completely blind, published *Paradise Lost.* It remains a bench mark in world literature. His works significantly impacted English thought.

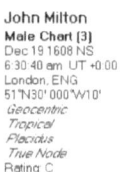

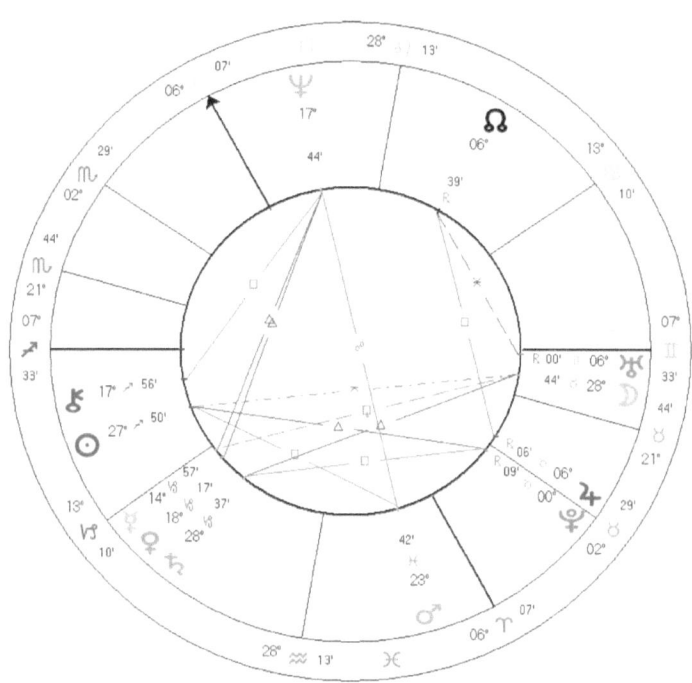

John Milton
Male Chart [3]
Dec 19 1608 NS
6 30 40 am UT +0 00
London, ENG
51°N30' 000°W10'
Geocentric
Tropical
Placidus
True Node
Rating C

Figure 20. Chart of John Milton

Milton's Chart, with the Sun and the Ascendant in Sagittarius, provokes a fundamental question around creativity—a process of building. Creativity, in his case, starts with the imagination. A look at the Second House of possessions shows Venus embraced by Mercury on one side and Saturn and the other. Mercury brings illumination and taps into beauty, Venus. Mercury manipulates thought and mental force into form, Saturn. One could additionally say that Milton radically transformed the past paradigm Moon/Uranus conjunction. The moon represents past forms, and Uranus, the planet of awakening and revolution, enable him to create a new dynamic in which we get a glimpse into both heaven and hell. As a Great Thinker, Milton held the potential to use the Venus/Mercury conjunction as a channel of inspiration. His mind was impressed with ideas relative to Universal Purpose, Pluto conjunct Jupiter in Taurus. Tremendous awareness infused the man/artist/writer/political thinker as he applied his mind to the improvement of the social conditions around him. Here the Pluto in Taurus influence is felt—the powers of transformation and illumination are activated in Taurus.

One of the fundamental urges of a soul with a strong Sagittarius emphasis is to align oneself with some worthwhile goal. Milton took on the articulation of the Great War in Heaven. Is there a nobler or grander preoccupation? Sagittarius brings in the 4th, 5th and 6th Rays. With all the zeal of the Sagittarius Sun and Ascendant thrust against the balancing act of the Libra Midheaven, we have a man caught between two worlds—Episcopal England and growing Puritanism. If ever a man was caught up in seeking a balance in the world, the Libra Midheaven, it was Milton. Much of his life he had to walk the tightrope, and he did it superbly. This was done with enormous personal turmoil and risk. In the end he went with the flow, and embraced Puritanism. This ushered in an era that witnessed the execution of Charles the First. Having sided with the Roundheads who eventually won a very bloody civil war, Milton must have felt that he made the correct choice. Later, when Charles II regained control, the poet had to re-balance his position and once more find his center.

All this conflict only served to stir up incredible powers of expression, and Milton is credited with giving the English language some of the greatest poetry ever conceived and written. The north node is in the Ninth House in Leo. He found success and satisfaction in expressing philosophical concepts in a dramatic manner. The Sabian symbol for the node placement, 10 Leo, sums up Milton's accomplishments.

"Early Morning Dew Sparkles As Sunlight Floods the Field. Keynote: *The exalted feeling that rises within the soul of the individual who has successfully passed through the long night which has tested his strength and his faith. Astrological Mandala*, Rudhyar, p137.

The North Node, place of ease and dispersion, signals the culmination of the creative process. Certainly Milton can be described as a reformer of thought, and in that sense he can be described as a transforming individual. The Keyword for this great poet is

Transfiguration. Once again we see in his chart the influence of the Venus. Here it is trine to Neptune. We have seen this configuration before in this discussion—Venus/Neptune, a magnificent entry for the 2nd Ray and additionally of Love through Beauty. Neptune is often called the higher octave of Venus, and when one considers the themes, subjects, and style of the verse, this is quite evident in Milton's expression of the Divine plan. Even in *Lycidas,* Milton displays talent that surpasses much of what has been written in the English language. He is credited with being one of the five greatest writers in English. With the flow of words, the great eternal themes and a point of view toward the eternal, he merits acclaim. The poem also suggests initiation of a sort. At the end of the long poem, the dead Lycidas appears as a resurrected figure. Light and the sun's rays aid his reappearance as he rises from the dead at dawn. Sabian Symbol for the Sun at 30 degrees Leo reads, "An unsealed letter. Keynote: *The realization that all thoughts and all messages are inevitably to be shared with all men.*"

As suggested by the symbol, John Milton's chart gives clues as to the integrated mind as well as the creative process. He often spoke of his muse. Divine inspiration came to Milton and he shared it willingly with his writings. Nothing, in other words, remained "sealed." Solar light pours through the First House of soul expression. The Sun conjuncts the Ascendant! Beauty of words initiates us into a world where God wars with Lucifer. It is a magical kingdom, where energies are in conflict and the yearning for a return is intensely stated. In the drama, Satan, at first dazed and confused, finds himself in the fiery lake of Hell. Eventually Lucifer pulls himself together and brings new, high voltage energy into the world. We must remember that Lucifer means "Lightbearer."

Later, the poet William Blake, along with many of the Romantic poets, insisted that Satan was the true hero of the poem. Satan eventually realizes his own dignity, and declares in incredibly solemn language his stance on freedom. It is doubtful that it was Milton's intent to glorify Lucifer, but that is how many readers have come to interpret the epic. This is interesting in terms of Esoteric thought. Esotericists equate the Fall of the Angels with the brave choice of incarnation that the Solar Angels made when they came to Earth. The mission was to stimulate Humanity's consciousness on the path of the Great Return. Over time, many have read into the masterpiece a hidden message revolving around the concept of revolt, integrity, self-expression and freedom. It is also worth noting the connection of Lucifer to Venus, for this is the name assigned to the planet when she appears as the Morning Star.

During the years when Milton's fame was growing, still another factor worked against him. During the mid-1640's, his eyesight began to seriously deteriorate. He always had bad eyesight as a youth, but in middle age the problem had accelerated. Milton found himself completely blind. The foreshadowing of this tragedy is found in the position of Chiron in the First House conjunct the Sun.

Milton suggests a new approach to God—one that is inner directed. His middle years coincided with the brief transition period during the reign of Charles I, Cromwell, and the Restoration of Charles II. After the Restoration, Milton spent a brief time in prison, but the popularity of his work protected him. He could have been hanged for his loyalty to the Commonwealth, but the elderly, and by that time blind poet, was spared. His latter years coincided with the Great Fire of London and the Plague. When *Paradise Lost* was finally published in 1671 it was well received. The great harmony between Mercury—intuition, and imagination—and Venus—beauty of expression—merges and blends with Saturn. The latter provides a potent combination of mental energies. Mercury brings in the Fourth Ray, and creates the correct connection for mental forms. Saturn brings in the Third Ray, producing patterns and structures whereby the inner world of form can manifest. Milton is dealing with Archetypal pattern. Even though brief in time, Puritanism, with its emphasis on individualism and faith, re-enforced Milton's characteristic belief in personal understanding and direct knowledge. Mercury, Venus, and Saturn collide in Capricorn. This stellium found resonance in Puritan ideals, and is reflected in Milton's stern moralism. Scorning the Episcopal Church, he associated himself with the Puritan cause and consciousness. Through contacting the core of his own being, the soul, he found his truth.

In this great epic we see the two sides of the poet—two sides of heroism. He is keenly aware of the presence of the Divine, yet he tries to create a scheme in which one can live a life in harmony with that Knowingness. There is once again the beautiful Neptune/Venus trine that has come up in so many of these charts. Neptune gives the flow of words the quality of Wisdom and the sense that true freedom is not external to one's experience. Freedom is a kind of release, a radiation from the core of ones life. Many of the treatises that Milton wrote dealt with freedom, and with that sharing he had an enormous influence on the thought of his time.

Fueled by a somber sense of Beauty, Goodness, and Truth, Milton's poetry conveys a quality of solemnity that makes us take the story and the magical weaving of it seriously. A sense of Beauty pervades the entire masterpiece. It is evident in the rolling free verse, the choice of words, the compelling rhythm—unsurpassed in the English language.

> Sing Heav'nly Muse, that on the secret top
> Of Oreb, or of Sinai, didst inspire
> That shepherd, who first taught the chosen seed,
> In the beginning, how the heav'ns and earth
> Rose out of Chaos . . .

In *Paradise Regained,* Milton introduces the idea of striving to regain what has been lost—a theme which has persisted in literature—showing how through continuous striving one can go past one's limits and mobilize energies in such a way that a return is possible. Adam and Eve have lost Eden, and Milton suggests that now they must strive after this fall. In this way Milton suggests a way to move into a new set of circumstances and events.

In his lifetime, Milton responded to a new group vibration. He wrote political treatises, pamphlets, and a very famous essay on divorce. Who would have believed that the frail, blind poet with his fervent belief in liberty and individualism would have a significant impact on the newly formed American colonies? In this way, Milton is another bridge builder between one cycle and another. He responded to the call of his time and has left a lasting legacy. He work is a benchmark of creative expression in human history.

It was his masterpiece, *Paradise Lost,* that was completed during the Venus transits. This magnificent work is a culmination of English poetry with its beautiful free verse and profound symbolism. Milton deals with existential symbols in words that are lush, provocative, and unforgettable. We must use this masterwork to gauge Milton's contribution, not only to the world of literature but also to the world of thought and the history consciousness. *Areopagitica* his defense of a free press, is echoed in the First Amendment of the U.S. Constitution. Milton died in 1674.

Charles I of England became King in 1625. Much of his reign occurred during the interim between the December transits of Venus. Charles was later executed on January 30, 1649. His public beheading, a marker event in the history of England, threw the country into turmoil. The English Civil War ensued. The Roundheads and Puritans took power in the name of Oliver Cromwell, but were later ousted when Charles II returned. Charles I was the only British monarch to be overthrown and beheaded.

1637:   In the interim of the transits, Ferdinand III of Germany followed in his father's footsteps and became the Holy Roman Emperor. His appointment of the Archduke of Styria was one of the triggers that led to the devastating 30 Years War in Europe, an extended event whose latter years coincided with the 17th century transits of Venus. This was a tumultuous period of history in a century marked by declared and undeclared wars. In a sense it was a time of religious war, as Roman Catholics, Calvinists, and Lutherans sought to prove their case. The Protestant Revolution of the 16th century, a momentous assault on tradition and standing authority, led to the anguish of ghastly wars. Chaos and uncertainty reigned.

Scientific revolutions led to the questioning of existing paradigms. The old model—the music of the spheres with the Earth at its center—gave way to the notion of the cosmos as a giant clock, a cosmos that could be measured, predicted, and comprehended. The Earth was no longer the center. Galileo had invented the telescope and discovered the transits of Venus.

Thought and cultures were changing. Still, at this time although there had been fundamental changes in values and institutions, Europe was fundamentally agrarian, with a handful of cities emerging as urban islands—areas where fresh ideas could concentrate, develop and eventually spill over to other regions.

The seeds of the Age of Enlightenment had been sown, and these urban islands would become the incubators as well as the distributors for the new ideas of reason, application of scientific principles, and revolution. Furthermore, the Divine Right of Kings had been directly challenged and the duly crowned king of England had been beheaded. A precedent had been set.

1761-1769 Transits

1759:   George Handel dies. Of Handel, Beethoven said, "Handel is the greatest composer who ever lived. I would bare my head and kneel at his grave." Handel's music is marked by its melodic boldness and dramatic sweep. He invented the English Oratorio, and was the supreme master of the Baroque. His passing left a void which would be filled by the surge of Romantic composers, of whom Beethoven would become the most eminent.

1761:   In 1760, George III became King of England. Named the 'Mad King George,' he was the ruler whose reign coincided with the American Revolution. The king was afflicted with a disease of the nervous system, a condition which greatly strained his ability to govern. Bouts with madness may have severely impaired his ability to deal effectively with the American colonies over taxation issues, and in the end the colonies declared independence in 1776. The final victory at Yorktown in 1781 led to peace and the treaty of Versailles in 1783. The loss of the colonies stretched George's sanity to the breaking point.

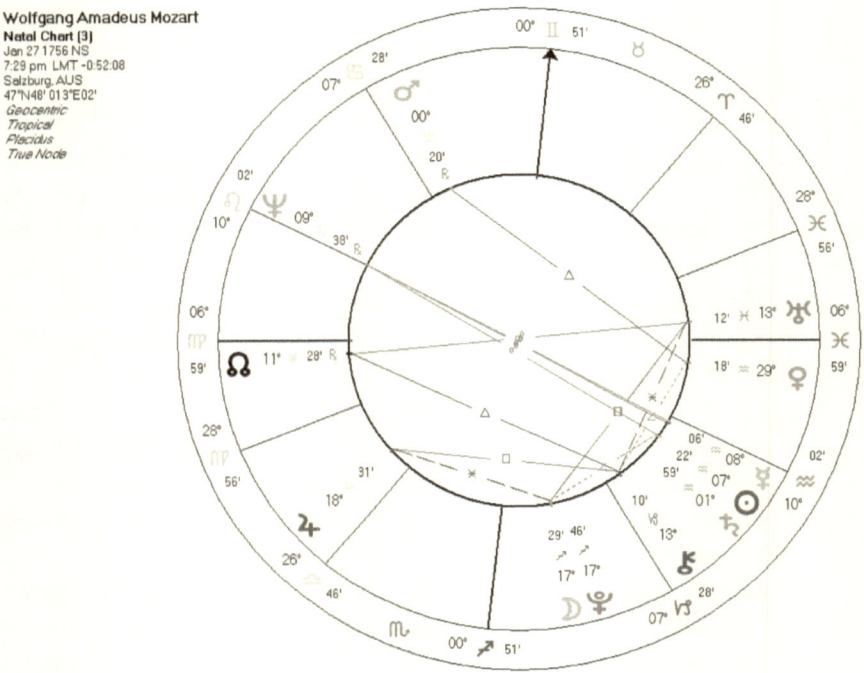

Wolfgang Amadeus Mozart
Natal Chart (3)
Jan 27 1756 NS
7:29 pm LMT -0:52:08
Salzburg, AUS
47°N48' 013°E02'
Geocentric
Tropical
Placidus
True Node

Figure 21. Chart of Amadeus Mozart

Amadeus Mozart was born in 1756. Shortly after the first transit of Venus, he achieved prominence as a well-recognized child prodigy with amazing keyboard skills. He was composing at the age of 5 and played before the Empress of Austria at the age of 6. As a child, he toured Europe and performed at the glittering courts in France and London. It is his compositions that set him apart as one of the greatest of musical talents. Hayden said of him, "Before God and as an honest man I tell you that your son is the greatest composer known to me either in person or by name. He has taste, and furthermore, the most profound knowledge of composition." With Saturn, the Sun and Mercury all in Aquarius in the Fifth House of creative output and children there is the indicator of one who can communicate in a structured, delightful way, but also one who could reach out to universal audiences. A Virgo Ascendant bestowed a vehicle for creativity. There is a sense of the Divine in his music. It also depicts a meticulous approach to his work. His Moon conjunct Pluto in Saggitarius bestowed idealism, ambition and intensity. Uranus squares both the Moon and the North node in the first house. There was bound to be a innovative break with the past. Pressure would be felt in personal issues. We know Mozart died a pauper and was buried in a pauper's grave. Mars trines Venus and in this magnificent artist's rendering of the new, lively and the beautiful we get a sense of the marriage of Venus and Mars. This was a life meant to communicate with Gemini at the Midheaven. Mars resides there bestowing much courage, vitality and joy. Intense, sensuous, Venusian qualities flowed in his musical creation that were not only refined but deeply moving. The Sabian symbol for the Sun, eight degrees Aquarius states "Beautifully gowned wax figures on display." The keynote is "Inspiration one may derive from the appearance of Exemplars who present to us the archetype of a new culture." Mozart most definitely collided with the fixed symbols of his time and manners of culture. He emerges as a visionary. He was able to create new forms and scintillating joyful ways of communicating through his music. The symbol suggests a sense of past as well as future expression. The wax gowns indicate a crystallized representation of ideal patterns. They are the kernel of what will be addressed and transformed. With the force of beauty and joy, Mozart strongly impacted the collective consciousness of his day. He celebrated the new, the joyful and provides a grand expression of the power of Venus.

Joseph Haydn was born in 1732. In 1759 he was appointed Vice-Kapellmeister to a leading Hungarian family and in 1766 Kapellmeister. As director of a musical group of between 15-20 musicians, he had enormous influence. At first he wrote instrumental music, but later expanded into larger compositions. He also devised an adventurous and coherent quartet form. He wrote over 750 works and 370 songs. His career was at hits peak during the 18th century transits.

The transits occurred during the reign of Louis XV of France. Under the rule of his predecessor, Louis XIV, Versailles had come to represent the highest expression of the Divine Right of Kings. Even though France had been at war for more than a century, Versailles was the most prestigious, imitated, and powerful court in the Western World.

But Louis XV was not to enjoy the success of this father— instead he was set to collide with prevailing chaotic and ominous forces. Trying in vain to overcome the mountain of inherited fiscal problems, Louis XV spent a good deal of time at Versailles, much of which was spent in the pursuit of women and the hunt. Famous love affairs with

Madame de Pompadour and the former prostitute Madame du Barry, along with other sex scandals, contributed to a decline in the king's popularity. Sex, as we have seen, like debt is a theme associated with Venus.

At the beginning of his reign he had been known as Le Bien-aime (the well-beloved) but by the end his reign the epithet had deteriorated to the well-hated. During his rule, French prestige suffered greatly. Unfortunately attempts to establish a more enlightened tax system were thwarted. Had this change been brought about, the fiscal calamity might have been avoided. To make things worse, the huge colonial empire was lost as a result of the Seven Year War—a humiliating cycle for the French monarchy. The stage for Revolution had been set.

It is important to note the relationship to social change and debt, a topic repeated in the theme of Venus, money and power.
This is a theme strongly present not only in the 18th century transit, but also today. Debt was a major problem in 18th century France and eventually brought down the monarchy. Similarly, debt is a problem today. We can, however, glean certain insights from the past.

1762: On September 22 of that year, the woman who we have come to know as Catherine the Great of Russia was crowned. Who would have believed that a German princess would become a ruler of Russia? Through a twist of fate she assumed leadership, ruling the nation as an absolute monarch in the tradition of Machiavelli. Catherine never lost a war and added extensively to the territory of Russia. The French called her Catherine le Grand, bestowing the masculine title in honor of her Martian-like feats.

Her Achilles heel? Love! If anything caused her trouble, it was cupid's arrow and the resulting love affairs that marked her as one of the world's great femme fatales. In many ways she was an enlightened ruler—Catherine instituted the reformation of the legal code, which had first been set in motion by Peter the Great. She may have been one of the first to advance women, or affirmative action as it is called today, when she appointed Princess Doskova as the Director of the Academy of Science. The appointment was made, not out of favoritism, but because the princess was well educated, intelligent and best suited to the job.

Even though Voltaire was born in the 17th century, his chart is included here because his work was widely accepted at the time of the 18th century transits and because he was widely known by the intellectuals of Europe. Voltaire spent some time at the court of Frederick the Great, a man who embraced the ideas of the Enlightenment.

With Uranus conjunct the Ascendant, Voltaire was a mover and shaker of the minds of men and women, radically impacting the values and mores of his time. The mere suggestion that the laws of nature could be applied to humanity and consequently determine the fate of men was revolutionary.

With Voltaire, and others, the Venus effect can be clearly seen in France of the 18th century where a group known as *the philosophes* gained visibility and influence. They were a group of public intellectuals coming from all disciplines of learning. They strongly emphasized the use of mind, of the intellect. Venus energy was evident in their application of reason to history, politics, economics, and especially social issues. *The philosphes* had a strong distrust of organized religion and state institutions. Many were deists. They strongly supported religious tolerance.

As the French Revolution came on the world stage, *the philosophes* faded into the background in Europe but left a lasting impact. The French Revolution reached a violent state in 1793. With the revolution came new thought forms and an unremitting desire for social change. The chaotic history of France endured for many decades and eventually played out with the defeat of Napoleon in 1812. This flow of history is mentioned to depict how the effect of reason and analysis can have tremendous effect in bringing about social change. Other factors are inclusive in the process, but here the power of mind and reason have been emphasized in the hope that we can learn from the cycles of history.

Descartes, Bacon, Copernicus, Kepler, and Newton had laid the foundation for the Enlightenment, many of whom did so at the time of the previous transit in the 17th century. Voltaire became a champion of systematic observation and experimentation. His confidence was aided and sometimes injured by a sharp wit—Mars conjuncts Mercury. He was quick to challenge traditional practices and customs, and question the basis of traditional social privileges. With Jupiter in Leo in the Fourth, Voltaire made his home in many places. He found a hearty welcome in Prussia, where Frederick the Great took enthusiastically to the new ideas that the freelance intellectuals shared with him. For a short time, Voltaire became a favorite of Louis XV who made him a member of the French Academy. He was a friend and confidante of John Locke and Isaac Newton.

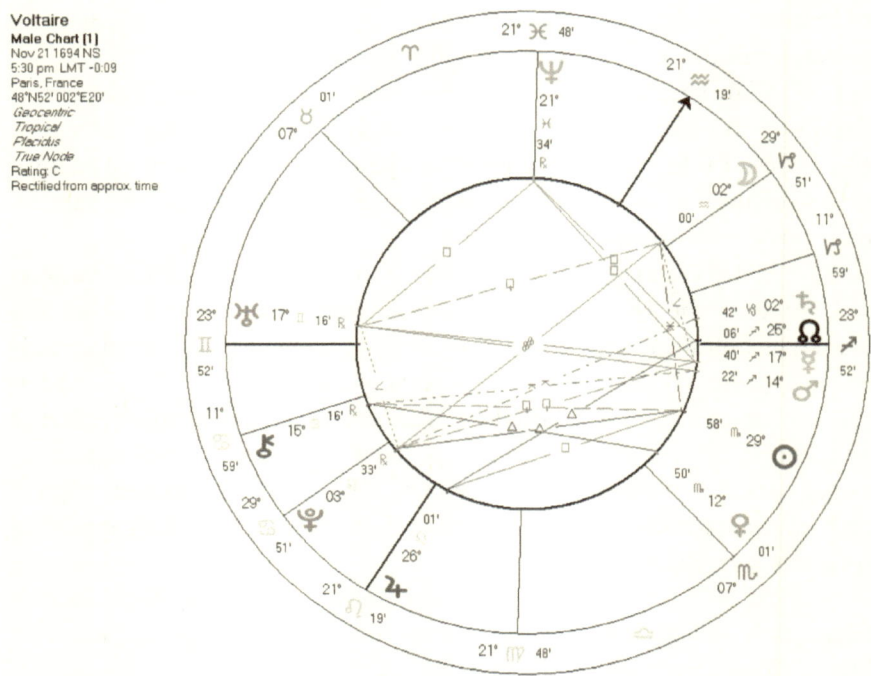

Figure 22. Chart of Voltaire

With his moon in the Eighth House, Voltaire amassed a fortune through careful investing and money lending. Though not a member of the aristocracy, whose privileges he scrutinized, Voltaire lived like one, having acquired a large estate where he hosted prominent intellectuals from all over Europe. The Moon position in the sign of Aquarius indicates a natural tendency toward humanitarian interests. Furthermore this placement suggests that there would be a natural inclination toward involvement in groups, preferably large groups, who could influence mankind. In addition, Uranus, the planet of freedom and liberty, conjuncts the Ascendant. No wonder he influenced American thought at the time of the revolution.

Voltaire died in 1778 having earned a reputation not only as an intellectual and champion of liberty, but also as an Encyclopedist and writer. His novel *"Candid"* combines acerbic wit with a compassionate view of human suffering.

Voltaire stands out as a giant among the thinkers. He influenced many people, and the combined effort of the prevailing time that created ideas and perspectives. These swept through Europe and set the stage for dramatic, sweeping changes.

It can be said that if only the ideals of the Age of Enlightenment had spread Eastward, the course of history would have played out very differently, but it is free will that drives the process and not the stars. It is our response to heavenly activity that is open to a variety of possibilities. And so it was that a vast majority of aristocrats who cherished their privileges were not so anxious to embrace ideas which would drastically change their position of preference and elitism. Such changes would have forced deep upheavals and caused them to pay taxes, give up serfs, and surrender monopolies on milling, baking, and brewing. With the rise of the middle class, fearful aristocrats held tenaciously to old traditions as they sought new ways to consolidate their position and enhance their wealth.

This reaction was especially true as one traveled East. Catherine the Great had tried to get the elite to relinquish their serfs, but it did not happen. The farther Eastward one traveled in the 18th century, the more entrenched was the notion of serfdom. While in France, the peasants owned nearly forty percent of the land in the East, the serfs were bound to the land, and the land was owned by the privileged class. In Russia, a serf could be bought and sold like property. This historical and sociological fact held Russia back while other countries were moving forward and outward. Human rights, property rights, the right to freedom and equality were unknown factors to the people of Russia at that time. This served as a long term drag on future progress for decades to come.

It can also be argued that since human rights were suppressed, Germany and Russia became a breeding ground for Romanticism. So it is that the energies that brought about significant changes in France and England played out differently in Germany and Russia. The upside of the argument is the artistic inheritance of the music, painting, and literature of the Romantic era that emerged from Germany and Russia. No one more than Beethoven stands out as an example of the period of rich artistic expression, the Age of Romanticism. For that reason, his chart has been selected as symbolic of that age.

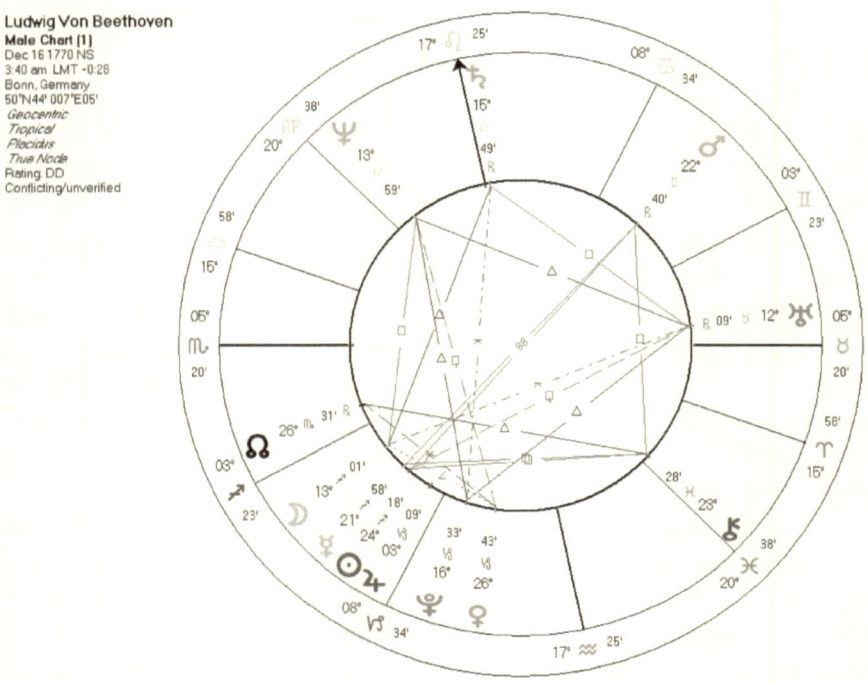

Ludwig Von Beethoven
Male Chart [1]
Dec 16 1770 NS
3:40 am LMT -0:28
Bonn, Germany
50°N44' 007°E05'
Geocentric
Tropical
Placidus
True Node
Rating DD
Conflicting/unverified

Figure 23. Chart of Beethoven

Beethoven's birth in 1770 ushered in an epoch of unforgettable music. No one more than he filled the vacuum that had been left with the passing of Handel. Beethoven's letters to his Immortal Beloved display an outburst of passion to an unknown person. The letters have given rise to huge speculation about the mystery person and her influence upon Beethoven's work. That he was still another giant is undisputed—along with Bach and Mozart, one of the all time great musicians and composers. His fascinating letters to his Immortal Beloved can be read on the internet. In a sense they can be read as if addressed to the goddess Venus herself.

Beethoven's chart reveals a Scorpio Ascendant. His music reaches from the depths to the heights of emotional experience earning for him the title of the great Romantic. With the Moon, Mercury and the Sun in Saggitarius we see the idealist, the striver, the prolific achiever. Like Mozart, he showed promised as a child but there was tension with the father—Saturn is in the Tenth conjunct the Midheaven. This Saturn placement also bestowed organizational ability and his work is highly detailed and classically beautiful although different from the past. Venus is in semi-square to the Moon suggesting problems with women. He had a pattern of loving women who were unattainable. Saturn squares Uranus in the Seventh. Relationships had to be difficult, upsetting and unique. Financial situations were charged as Mars is in the Eighth opposing the Sun and Mercury. Jupiter resides in the Second so there is a promise of success with a conjunction to the Sun. Neptune trines Pluto and Uranus suggesting a harmonious fusion of an ideal goal and the spiritual dimension. Idealism and structure blend into something beautiful. There are hints of the social transformations occurring all around him in the expression of the music. Mars opposes both the Sun and Mercury. Problems with communication had to surface and we know that Beethoven eventually became deaf. The Sun conjuncts Mercury suggesting not only the creative impulse but also the cohesive power of love. The Sun Squaring Neptune may at times have caused confusion of Soul purpose to that of the personality. There was the ability to appreciate the inclusive quality of life.

This is a life that required periods of solitude and quiet. The Ninth Symphony with its Ode to Joy signifies a triumph of Soul over the mundane. Here the grand trine is apparent and triumphs—Pluto, transformation, Uranus, change, and Neptune the spirit of music itself. With Venus in Capricorn here is an artist ever at work creating structures and interactions between people nations and goodwill. One can say that Beethoven's music, in the end, sought to bring a cohesive quality to all who heard it. Beethoven's music penetrates to the core of humanity. Each listener responds in his or her own way to his genius.

The universality in Beethoven's music gives insight into Venus and her ways. She can harmonize the diverse, she can bring about more creative and pleasant outcomes, while Mercury will awaken consciousness. All this is evident in the creative output of Ludwig Von Beethoven, a composer who forever changed the world of music.

Venus directs the spirit toward the beautiful. Down through the ages, her form has been painted repeatedly, been invoked by words or the music of Hayden, Mozart and Beethoven.

In the world of art, no one can deny the great tidal wave of Venus paintings that were created after the 18th century transit. Painters like Ingres, Fragonard, and Boucher wrought magic with their brushes. Morning glory blues, light streaming in pale yellow, unsurpassed flesh tones, and luminous yet life-like figures made it clear that the femme fatale of Olympus was worth adoring. These paintings display the body sensuously delighted in the expression of love. This acceptance and celebration of the feminine in such glorious light seemed to reconcile the sensuous, and all that related to the body, with the transcendent-spiritual realm.

The 18th century transits in a very real sense stimulated the creation of the intellectual and cultural challenges that would culminate in the French Revolution later in that century. Enlightenment thinkers exalted rational and scientific thought, a process that they believed would reveal the secrets of the universe. This universe would reveal itself as orderly, and operating on a set of natural laws. Meditation, religion, or divine revelation were not necessary to penetrate into the truth of the Cosmos. This movement challenged traditional beliefs and questioned the ability of the church to be the intermediary between God and His people. The Age of Reason profoundly affected the rise of secularism. John Locke went so far as to suggest that government is a contract that must protect certain natural rights to life, liberty, and property. In Prussia and Austria, ruling monarch's practiced "enlightened absolutism." Frederick the Great of Prussia saw himself as the servant of the state. Contrast that notion to Louis XIV's statement, *"L'etat c'est moi,"* I am the state. Of all the states of Europe, it was the French monarchy that most strenuously resisted the ideas of the enlightenment.

The point of this discussion is that the confluence of these issues and events point to the 18th century transits of Venus as a time when channels of intellectual communication shifted away from dynastic leadership and a court focus toward that of educated groups of individuals who moved about Europe, spreading ideas of optimism and greater inclusivity. Voltaire's chart has been cited for he stands out as one of the free-lance *philosophes*. These individuals wrote to each other and significantly influenced the tide of events. The population was growing due to significant changes in agriculture, warfare, and even health care—though medical treatment then would horrify anyone today—to the extent that a more educated and just society emerged. These men and women read newspapers, journals, pamphlets, and books. The economies of France and England had improved, creating a larger, prospering, and less old-fashioned middle class. The challenges that arose around the time of the 18th century transits would, along with the reactionary stance of the French monarchy, develop into a multi-pronged assault on the *Ancien Regime*. The French Revolution exploded in 1789.

Beethoven's life and work demonstrates how the artistic Fourth Ray mind can work the themes of repression and exaltation into great expressions of suffering, freedom, and triumph. On the personal level Beethoven faced poverty, melancholy, and failure in love, experiences that when linked to a larger group consciousness found expression—his unforgettable music remains a tremendous legacy.

Peering backward through the lens of time one can see the legacy of Venus in the history. By the end of the 18th century, the influence of the new *philosphes* had been felt in the new world. There, the founders of the United States were greatly influenced by the emphasis on reason, religious tolerance as well as the concept of Deism. The constitution of the United States bears the mark of Venus and *the philosphes* of France.

In Europe urbanization increased. This move to cities deeply restructuring the foundations of culture and society. Beautiful music that created new forms and uplifting and romantic themes stirred public awareness. Kingdoms fell—new nations emerged.

1874-1882 Transits

Time marches forward, and a new set of transits appear. During that period, 1874 marks the birth of Winston Churchill. More than any one individual, his courage and vision saved the western world from the darkness and horror of the Nazi regime. He took over the reins of leadership during World War II, and returned again at the age of 88 to fashion a post-war Britain. Shortly before his death, the newly elected president of the United States, John F. Kennedy, pronounced Churchill an honorary American.

Churchill's Libra Ascendant speaks of one who is a peacemaker. With Mars conjunct Jupiter in the first house and in Libra, this in a man who had the ability to fight, and to fight an expanded, in this case, world-wide campaign in defense of his values. Cancer on the Midheaven relates to dealing with the masses, and Capricorn at the Descendant brings structure to what needs to be done in respect to the Motherland.

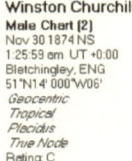

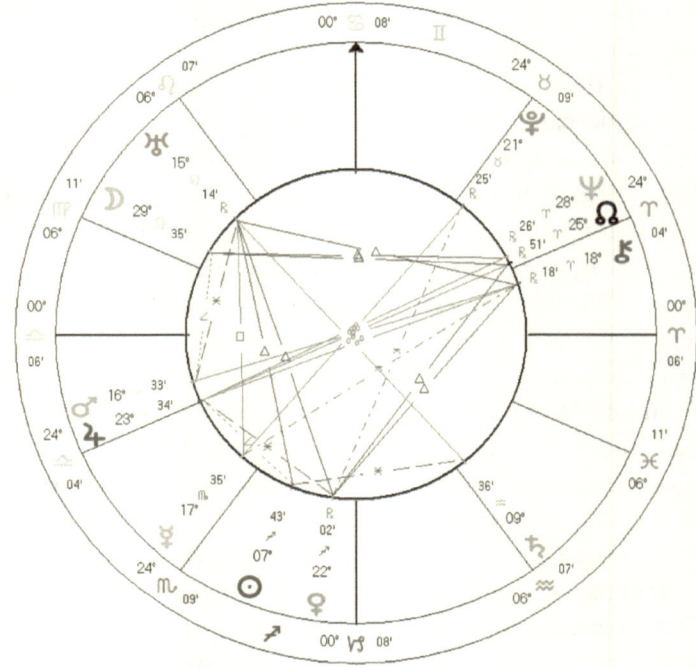

Figure 24. Chart of Winston Churchill

For purposes of this discussion, notice that Venus conjuncts the Descendent, implying a great love and respect for his homeland, especially in time of wounding. Chiron trines Venus and provides the bridge out of suffering. Like Christ's willingness to sacrifice, the British people took on much of the burden in the early days of WWII They held on to their culture and to their civilization with a ferocity that was heroic. With Mars in the First House, Churchill was able to engage that heroism and take a stand. The outcome and historical maker is that England is credited with saving the West from Nazism and a dark age.

Pluto opposes Mercury, which energized Churchill into a powerful speaker and writer. He won the Nobel Prize for his history of the Second World War, in which his eloquent words provided a vehicle for the expression of Will, couched in exquisite language. Venus trines Uranus—a condition which brought unusual light, clarity and feeling into the situation. With Venus conjunct the DC—a royal chart, a chart of destiny—we see the pattern of one who had the potential to do the work when it was called for. Neptune opposes Jupiter, with the potential of enormous 2nd Ray influence.

This is a chart strongly influenced by Venus. In his later years, Churchill allowed himself the pleasure of watercolor painting, a hobby that took him to a professional level. It is worth noting that his delicate, fluid watercolors display a personality whose soul sought expression in refinement and beauty.

1874 Transit: Susan B. Anthony's 1873 speech set off a whirlwind of events that only increased in force. She gave a speech in favor of women's right to vote, despite having been indicted for what had been considered illegal voting in the previous presidential election. Susan maintained that she had not broken the law, and should be protected by the Constitution of the United States of America. Susan B. Anthony was one of the strongest advocates of Women's right to vote in the mid-19th century. She is representative of a whole host of women who struggled—first for the abolition of slavery, and then for the right of all citizens to vote. This commitment was unwavering, even after the slaves were freed and black men eventually got the right to vote. With the exclusion, that is the fact that although women abolitionists worked to achieve citizenship and the right to vote for former slaves, women themselves did not achieve the right to vote for some time. They were excluded.

For this reason, Susan and her cause suffered a bitter setback, but she and her group of women advocates never wavered in the vision of universal suffrage. Women finally won right to vote in the United States in 1920.

With a stellium in the Tenth House, Susan B. Anthony was bound to play a prominent role on the world stage. All of the planets except one, Mars, are above the horizon—another expression of outward direction. All but two of the planets are in the East, indicating that here is an individual who was driven by inner certitude.

With Pluto, Venus, Chiron, and Saturn in Pisces, we can say that she was committed to serving within the constraints of a Piscean field, taking action for the emancipation of the slaves and for the transformation of women's role in society.

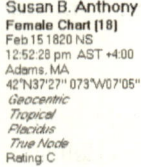

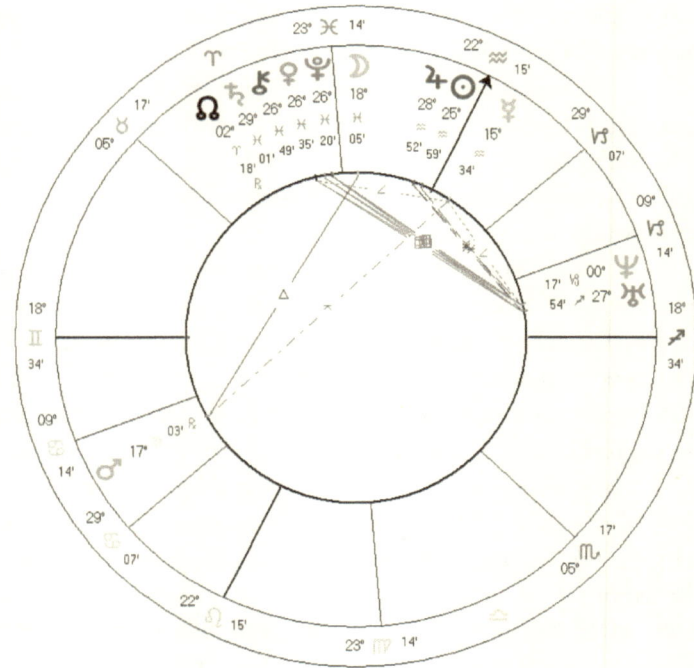

Figure 25. Chart of Susan B. Anthony

Anthony's Venus/Pluto conjunction points to her ability to support a deep transformational change in terms of culture as it relates particularly to women. This kind of focus and committed attention endured during her entire life. No wonder she is called the "mother of us all." Venus is exalted in Pisces, indicating Anthony's ability to dissolve the separatism inherent in her life work and eventually bring about integration. Discrimination is sharpened, with the urge to love all universally.

With a Gemini Ascendant, Susan B. Anthony can be called a teacher of love and acceptance. Venus, the esoteric ruler of Gemini, influences the teaching faculty. The Sun, Jupiter, and Mercury form an impressive triad in Aquarius. Mercury in this position intensely relates the mind to archetypal patterns relative to New Age principles. Jupiter in close vicinity only amplifies the 2nd ray of Love, and Jupiter is naturally associated with Aquarius, giving even greater energy to this placement. Such a person seeks to unite groups within groups. With Susan we see how she fought, first as an Abolitionist, and then for the Emancipation and the Negro right to vote, and then for Women's suffrage. With the exclusion came the right for the freed male slaves to vote, but all women were excluded. This brought great disappointment to her cause since the 13th amendment passed and she could not yet rally sufficient support to win the vote for women. Suffrage never happened in her lifetime, but the Saturn persistence in the 10th House never let up. A very active vehicle for the expression of Love/Wisdom, the Uranus/Neptune in the 7th House speaks to unconventional relationships and to radicalizing relationships in general. The call to service was immensely strong. Susan B. Anthony stands as another bridging individual—her effect on culture and civilization is still being felt. The Sabian Symbol for her Sun is 26 Aquarius. "A Garage man Testing A Car's Battery With A Hydrometer. Keynote: *Skill in applying natural laws to the solution of everyday problems resulting from life in our technological society.*"

Inventive genius is clearly indicated. Another major theme is the need to manage the complexity of interpersonal, intercultural, and political issues of society. Here, a symbol of mental efficiency is evident, and reflects the kind of life of one of those people who well use their intellectual powers of observation and analysis to check on intended results. We know that Venus stimulates keen analytical thinking and from Anthony's life, we can extrapolate how the mind and reason will play out in the future in a more general sense.

Still another major figure of the time is Claude Monet. He was born in 1840. To escape the Franco-Prussian war, Monet fled with another painter, Pissarro, to London. There they studied the works of the great English painter Turner, a romantic painter known for his use of light. Monet was deeply affected by Turner's approach and incorporated many of the unusal techniques he learned during his stay in England. In the end, Monet created a whole new approach to painting and revolutionized the art form.

In 1874, he painted *"An Impression Sunrise."* This painting gave the name to the first exhibit, Impressionism. The name stuck, and is used to refer to a whole group of painters including Gauguin, Van Gogh, Manet, Monet, Pissarro, Sisley, and Mary Cassat. The group, centered in France, was active between 1860 and 1880, but it is notable that the first show under the title Impressionism coincides with the first 19th century transit of Venus.

Monet's impressionistic style employed an innovative approach that captured the fleeting moment. It was as if he could freeze in time any given moment. These paintings were bathed in fresh colors and gave rise to a whole new style of out of doors work called en plein air. Light and its effect on color incorporated vivid hues and gorgeous landscapes. Little points of light were rendered in such a way as to instill a kind of livingness never before seen. Monet persists, even today, as one of the most loved and appreciated of artists. His influence lingers, his style is copied.

Degas, born 1834—1917: This remarkable painter is usually associated with the impressionists probably because of his use of pastels and his fluid style. Because of his classical background, style, and subject matter, it can be said of him that he was the most modern of painters, with his natural and spontaneous depictions of ballet dancers. These paintings, many of which were created between the 1874-1882 transits, express beauty and form in such a unique and splendid fashion that Degas became a major contributor to art in general. Who can forget Degas' pictures of the ballet, so realistic and bursting in color?

1880: Four years after the first 19th century transit, the London Museum of Natural History opened. This truly remarkable institution not only fostered research in natural history but also became a forum where the public could learn and see the wonders of natural structures within form. It was novel in its time, and contributed greatly to the expansion of consciousness in the area of science and concrete knowledge. This institution provided a forum for things new and original as well as scientific and beautiful.

The famous poet, Arthur Rimbaud, was born in 1841. Known as the boy wonder of French symbolism, he contributed some of the most remarkable poetry of the time. His work began to be translated into English during the 19th century transits of Venus. Drawing heavily on the subconscious, the astral world was no stranger to him. Most important, in terms of innovation and new trends, he is identified as one of the creators of free verse a form greatly used even and especially today.

An example of his style sheds light not only on his work, but also on the innovation that colored the 19th century transits. In a way, it is a summation of the kind of new, fresh turns and twists that can happen at these remarkable times. As he said:

> *I am the inventor more deserving far than all those*
> *who have preceded me; a musician, moreover, who has*
> *discovered something like the key of love . . . from Life.*

The effect and influence of Venus and also of Psyche is evident in Rimbaud's work and aesthetic vision—the key of love . . . from Life. This phrase clearly links the power of love to the most essential, powerful will to be, life.

His daring validation of his point of view and his style ideas demonstrates and validates his need to break out and create something new and ever more beautiful. He created new forms. Victor Hugo, another transforming writer of the time called Rimbaud "an infant Shakespeare."

These impulses are the kinds of things that can happen when Venus transits the Sun. This final quote summarizes Rimbaud's originality and tone.

> *I have stretched ropes from steeple to steeple;*
> *garlands from window to window; golden chains from*
> *star to star, and I dance. Rimbaud*

On a more somber note it must be stated that Rimbaud provides insight into what can happen when a person's "equipment" is not sturdy enough to integrate the incoming inspiration. Indeed, his ropes were stretched. Rimbaud was known as one of the poets of the "decadent" period. This is because of personality issues as well as for his attraction to drugs. Here is an example of one whose mind-body-emotions could not handle the influx of light. He died before he was thirty years old. His career peeked and he died leaving a remarkable legacy of imagery, beautiful language and other worldly expression.

Still another notation for these past set of cycles is the year 1882—Charles Darwin dies. Some have credited him as having made the greatest contribution to the history of man, science, and of consciousness. Having published *On the Origin of Species*, he became the eye in a storm and faced a lifetime of criticism, controversy, and skepticism. In 1871 he published *The Descent of Man*, a work which synthesized his findings and his position. By the time of his death, his ideas had been largely accepted by the scientific community of the time and his theory prevailed.

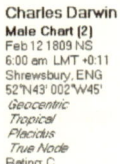

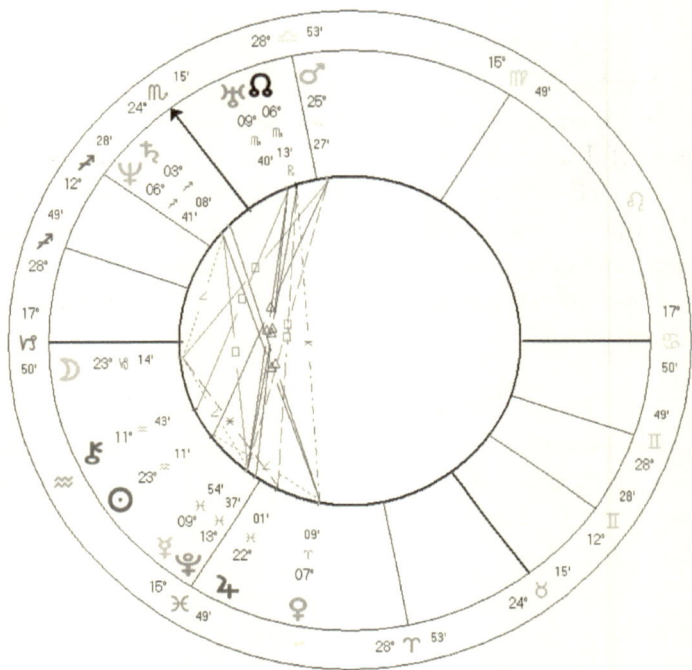

Figure 26. Chart of Charles Darwin

Darwin and scientists like him did not provide all the answers, but they did open up new directions. Moreover, he gave us a method to use whereby further discoveries and definitions might be made. We know so much more now than we did at the first transit of Venus in the 17th century. Laws have been discovered; new and penetrating forms of art have been created. Revolutions have occurred and whole continents have been transformed. We still ponder upon the purpose of all that we see about us, still wonder at the mystery, and question the destiny of the human race. Darwin opened new pathways to learning, to the scientific approach and to a deeper understanding of life on the planet. He was a transformative figure in every way, a revolutionary.

Darwin had his Sun close to the nadir and in the sign Aquarius.

His vision was fresh, new and innovative stemming as it did from a deep conviction and foundation in the scientific method. The Sabian symbol for the Sun is incredibly accurate. "A Man, having overcome his passions teaches deep wisdom in terms of his experience." The keynote is: "The constructive use to which difficult past experiences can be put as examples for those who are still striving to overcome their passions."

Darwin was a fanatical naturalist. His experiences and resulting conclusions led him to upset existing paradigms in a bold step to explain the natural world. His theories of evolution caused deep rifts in the scientific community. The idea of natural selection caused a revolution in the biological sciences. He wrote the **Origin of the Species** in 1850 and the **Ascent of Man** in 1879.

> It is not the strongest of the species that survives, nor the most intelligent that survives. It is the one that is the most adaptable to change.
>
> Charles Darwin

Words like this earned him not only fame but also contention. Yet he persisted and continued to press on for the efficacy of the scientific method.

> Ignorance more frequently begets confidence than does knowledge: it is those who know little, and not those who know much, who so positively assert that this or that problem will never be solved by science.
>
> Charles Darwin

In a way, Davwin served as a mentor assisting the world at large to take a step forward in growth, in evolution. Here is a true educator. Moon in Capricorn allowed him a secure classical structure from which to work and reach out to new interpretations. Once again we see a Sun/Venus connection in the trine. Venus bestows analytical powers. Perhaps, not since Aristotle, has there been a mind so penetrating, determined and fresh.

Working out his own theory Darwin imposed an Aquarian perspective into a Piscean world. Mercury, Pluto Jupiter all reside in

Pisces. We should not be surprised that Darwin's theory caused turmoil, anger and even disgust and fear. His Mars in the tenth house lent courage to overcome the obstacles of past paradigms and thought—religion, science and philosophy were impacted, turned upside down, and eventually revaluated, all because of his theory. There was a good amount of courage and the daring to see it through. Moon in Capricorn squares Mars in Libra. The Libran influence can quiet the potential for rashness but the courage remains. Neptune and Saturn conjunct the Ascendant in Saggitarius. Here was an idealist, one who aimed high. His task to confront the old and establish what is new, threaten the existing order especially the depths of Christianity. Pluto in Pisces, the sign of Christianity conjuncts Mercury. This powerful Pluto placement gave Darwin the ability to communicate the new and to shatter old crystallized forms. Mercury leads Pluto, an aspect which somewhat eased the situation. The ideals of an unfolding evolution were radical and gave a new dimension to spirituality and a creation on the path of growth and development. An yet, we see once again how the new is built upon the old.

The ideas of Plato and Aristotle are embedded in Darwin's "Origin of the Species." Similarly, there is a strong correlation with an esoteric view of life. In speaking of natural selection Darwin writes:

> There is grandeur in this view of life, with its several powers, having been originally breathed by the Creator into a few forms or into one; and that, whilst this planet has gone cycling on according to the fixed law of gravity, from so simple a beginning endless forms and most beautiful and most wonderful have been, and are being evolved.
> Page 243 Origin of the Species.

We can compare Darwin's point of view with that of the esotericist who holds that the Logos works out his purpose through a plan in which the infusion of soul light enlivens and elevates all structures in their path of evolution. Much more on this topic can be learned from Alice Bailey's "Treatise on White Magic"—a foundational document for all students of esotericism which explains in detail the process of evolution. Without a doubt, Darwin symbolizes the use of analytical mind, a refined sensitivity and the ability to endure. He is a major figurehead to be associated with the Transits of Venus not only becasue of his scientific acumen but also because of his philosophical contribution.

It should come as no surprise that Adolph Hitler was born during the 19th century transits of Venus and at a time when the German people were deeply involved in uncovering and rediscovering their ancient past. Indeed it was the Germans who invented the concept and word, *Kultur,* Culture. Under Hitler's leadership, Germany struggled to create a new civilization, not only for themselves but for the whole of Europe. Hitler saw himself as a man of destiny, accentuated by the moon in Cancer. Saturn at the Midheaven gave him the strength and endurance to forge ahead as the leader of the "master race" as it was termed. He assumed the title of Fuhrur and his ignominious dictatorship lasting for only thirteen years.

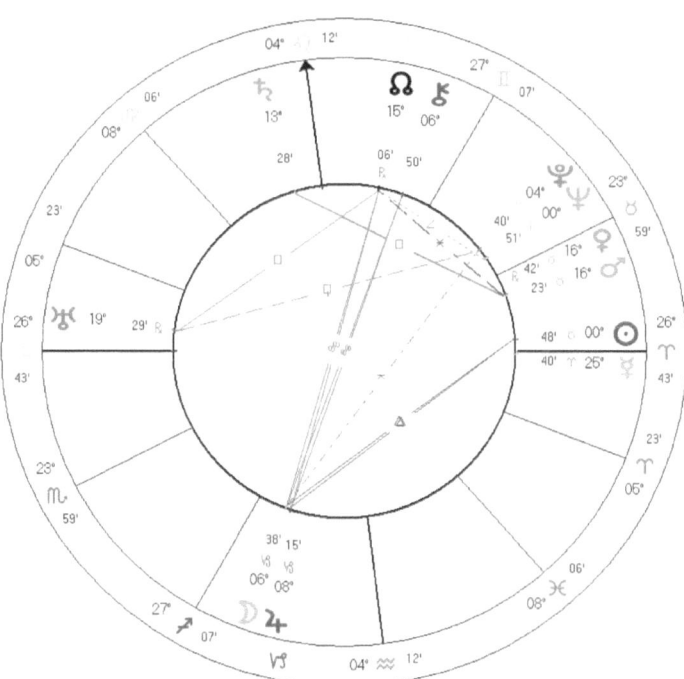

Adolf Hitler
Male Chart (1)
Apr 20 1889 NS
6 30 pm LMT -0.52
Braunau. Austria
48°N15' 013°E02'
Geocentric
Tropical
Placidus
True Node
Rating: AA
Quoted BC/BR

Figure 27. Chart of Adolph Hitler

Hitler and his creation, the Third Reich, deeply impacted not only world politics, but the culture and consciousness of Humanity on a world-wide scale. We can deduce that Nazism and Hitler as its leader produced a major paradigm shift in the latter half of the 20th century. Uranus at the Ascendant and in Libra promised a major shift in the balance in the nature of things. The shift was radical, electrifying and devastating.

Hitler's policy of Aryan supremacy fell on accepting ears. Germany was still reeling from its defeat in World War One and was eager to find a cause to uplift not only a depressed economy, but also a depressed national psyche. Hitler's Sun conjuncting Mercury with a trine to both Jupiter and the Moon, endowed him with magnetic powers of speech and persuasion. The power of death and the wielding of that power is expressed in the Eighth House—Pluto conjunct Neptune trines Uranus—coloring his consciousness with radical fanaticism. Hitler was the author of one of the darkest feats of human history—the holocaust and the death of millions. Germany's defeat by the allies and Hitler's death signaled the death of the old Europe.

What is important for this discussion and helps to unravel the latent secrets is the synthesis of themes and motifs that we can glean from these most recent transits of Venus—the confluence of ideas and themes out of which emerged the world of the 21st century.

Darwin had transformed the natural sciences forever, and by the time the transits occurred, his life's work has reached a crescendo that received public acclaim and acceptance. The Women's Movement and Suffrage had been set on a solid path. Susan B. Anthony has left her mark. Art, music, and literature reached out to a freer, unbounded expression that transformed the old forms into new, fresh concepts. Hitler's Germany and his idea of a United Europe took a strange twist of fate. A defeated Germany arose from the ashes and became a major player in the creation of the new European Union. The persecution of the Jews encouraged the ultimate creation of the State of Israel. These relationships and corresponding conditions provide the immediate problems and challenges that shape the world today, and provide hints into the kinds of social and cultural changes that the 21st century transits herald.

It must be remembered that the heavens do not necessarily foretell events. The transits of Venus represent a window of opportunity during which radiant and electrifying energies shower down on earth.

How we respond, and to what degree these incoming cosmic seeds impact our thoughts and our relationships, depends upon how we nourish and attend to them. We reach up to the stars with imagination, hope, and fervor. The heavens for two brief transits infuse us with the gifts of Venus, Gemini, and the Sun. During the last completed transits of Venus, Europe reached the zenith of global power. In the century that followed, all that would be changed.

Before leaving the cluster of individuals who lived during the 19th century transits, we must consider the life and the astrology of yet another remarkable human being—Albert Einstein. Born in 1879, Einstein challenged prevailing paradigms about the nature of reality, of matter and of energy.

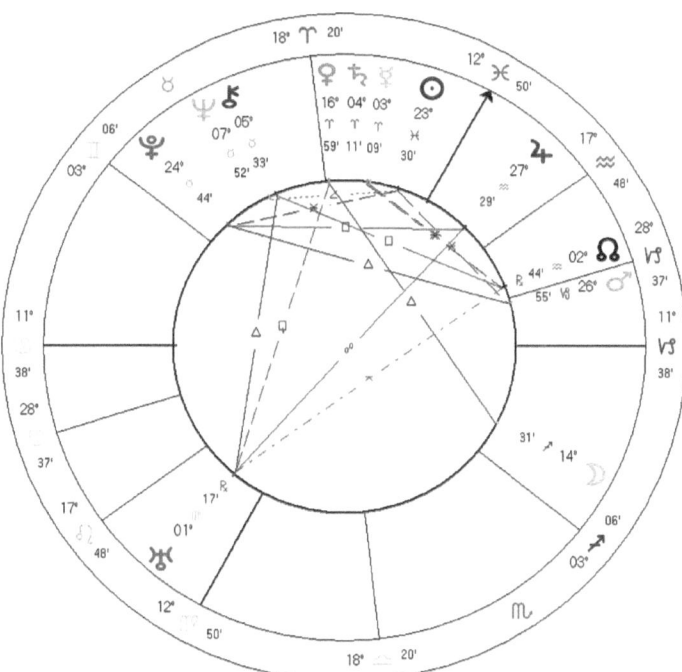

Figure 28. Chart of Albert Einstein

He is known to have said, "Great spirits have often encountered violent opposition from weak minds." His Sun in the 10th house in the sign of Pisces allowed him to see through existing barriers, and to immerse himself in a life long study of existence itself. The Sabian Symbol for the Sun at 24 degrees Pisces is, "On a small island surrounded by the vast expanse of the sea, people are seen living in close interaction. The keynote is, "The need to consciously accept one's own personal limitations in order to concentrate ones energies ad live a centered and fulfilled life." Einstein's early life was difficult. He was known to have an unusually large head and this no doubt encouraged teasing. He was a rebel from the beginning and was expelled from school. He must have resolved the personal limitations early enough to begin his major work while employed in a patent office.

Pluto squares Jupiter with the force of greatness and transformation—he confronted fundamental issues about life, light and the cosmos. A stellium of Venus, Saturn and Mercury in Aries in the 10th House suggests not only mental brilliance but the needed analytical skill to penetrate into the very nature of the cosmos, gravity and magnetism. Einstein won the Nobel Peace Prize in 1922 for his work on relativity. This is reflected with Jupiter in the Ninth House of higher mind. The north node in the Eight House of resources and other people's money indicates some fulfillment from other's money. This is evidenced in winning the Nobel Prize. Jupiter opposes the Third house Uranus suggesting the cutting edge thinker in a relationship of tension. Venus is Aires only serves to enhance the mental brilliance.

"The eternal mystery of the universe is it's comprehensibility." Einstein.

Uranus opposes Jupiter and it is no wonder that he broke through old paradigms in a way so devastating that nothing would ever be the same. His discoveries led to the development of the atomic bomb, the single most destructive weapon of war ever devised. At first he supported the development of the A bomb as he knew others would be at work to do so. But later, Einstein, along with the philosopher, Bertrand Russell, came out strongly against the use of the power of fission as a weapon of mass destruction. Neptune in the Eleventh House (institutions) trines Uranus in the fourth (his roots, his core.) His higher self and connection with humanity had to rebel and Einstein became well known for his avid pacifism.

Venus trines the Moon in Sagittarius so there is an elevated kind of idealism stemming from past history. He was offered the presidency of then new country of Israel but turned it down. Mars in Capricorn in the Seventh house indicates an ability to demonstrate courage in relationships as well as the kind of turmoil that resulted from upsetting old ideas about the light, the cosmos, and magnetism. Einstein prepared the way for the emergence of quantum mechanics and in so doing turned the world of science upside down. Writing hundreds of papers, he opened new door ways for future study. Uranus in the Third house indicates an unusual and cutting edge thinker. He is most known for his Special Theory of Relativity, $E = MC2$.

His belief in the mind to discover the truth resonates with Venus. He is known to have said the following.

> All that is valuable in human society depends upon the opportunity for development accorded the individual.

Einstein's belief in the mind as a tool to naturally discover and use the truth resonates with Darwin. In both cases they held to the idea that reality could be known, understood and applied to the world of form. These were core beliefs for both men, revolutionaries in thought, science and cultures.

With Mars in the Seventh and in Capricorn, the structure of his personal life was chaotic. Some credit the foundation of his early work to his wife who was said to be a brilliant mathematician in her own right. With Mars in the Seventh, turmoil in personal relationships could be expected.

Einstein remains an icon of the age. He is regarded as the father of modern physics and one of the most notable intellects to have ever lived. Known to master detail, analysis and discrimination, a capacity to think and act scientifically he also had the power to discover and verify through experimentation. These are qualities that he shares with Aristotle and Darwin. All three reveal the secrets and opportunities latent in the energy of Venus.

Chapter Nine

ON THE BANKS OF THE EUPHRATES

Civilization and culture, those two areas within the domain of Venus, can be seen as the product of many interconnected relationships. Friendship, religion, rivalry, enmity, chance encounters, kinship bonds, economic exchange, military conquest, and ecological impacts—these factors interact, and people devise ways to live within the frames that are created. Planetary influences converge and a model for what describes economic conditions can be expressed.

We might consider a triangle of the following:

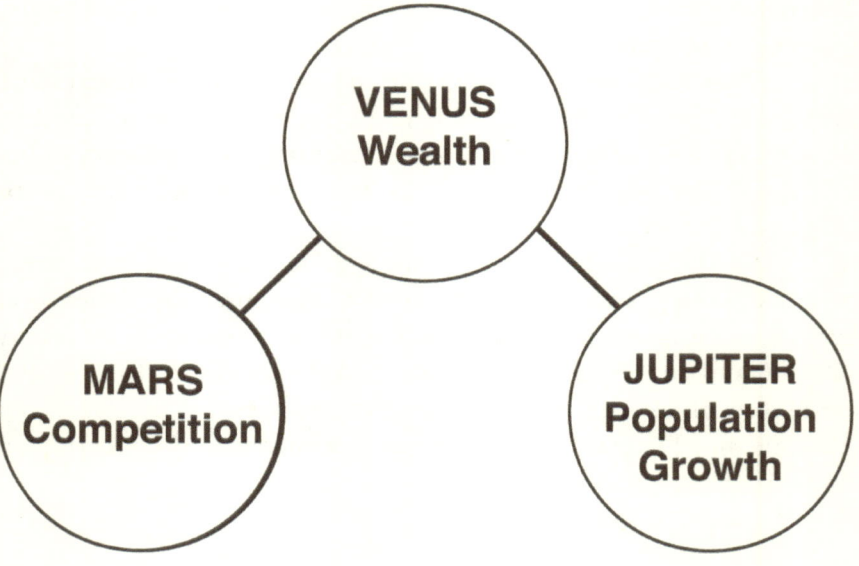

Figure 29. Economic Factors

The impulses from each of the three points of the triangle communicate with each other, and language, money, and writing developed. Within these connections, people exchanged technologies, religious ideas, and inventions; they incorporated or they sloughed off useful ideas. Inadvertently they exchanged diseases. The latter destroyed some and made others strong—in other words, to quote Darwin, "the survival of the fittest."

As mentioned earlier, Venus is what drives people toward their desire. On the higher turn of the spiral, it is what drives us toward what we aspire. The transgenic code that we all carry bestows the gift of language as well as the gift of building—architecture. Within this frame, Venus works in concert with the devas, the great builders and maintainers of form, in such a way that these tendencies to relate unfold naturally in a human being. Life evolved as Darwin asserted. People clustered into groups and groups into larger groups until organized and even larger groups emerged.

Civilization as we know it began on the banks of the Euphrates River nearly 6,000 years ago, with the cities of what we call ancient Sumer. City life began to grow because people realized that there were certain advantages to cooperation and the division of labor. Military advantage was no small boon, and it could be gained as larger groups of people could afford to staff full time, well-provisioned soldiers to protect them and to wage war. Coins were minted and a language of mercantilism developed. Writing was a practical short cut to what in the past had been a complicated method of bartering things and ideas. Where oxen had been used as a means of exchange, coins in the shape of the bull's horns were made to facilitate the exchange. Venus via Taurus made her influence known! Of course some people remained outside the metropolitan regions, but the trend for urbanization had begun.

From this early web of connections that became known as Mesopotamia there developed the 20th century invention—Iraq. The arbitrary boundaries that led to the creation of this Middle Eastern country are important for this discussion because it remains one of the major flashpoints on the planet today. Since its birth, Iraq has lived an extremely precarious life. Its modern day history deserves some analysis in light of recent events.

World War I in the 20th century, led to new interest in the desert regions of Northern Africa and the Middle East. This land had been ruled for over four hundred years by the Ottoman Empire, who had become a war ally of Germany. The Ottomans had clustered the tribes into general areas around Mosul, Baghdad, and Basara. Eventually, these groupings of people would later become what we today call Iraq.

The now famous Lawrence of Arabia was sent to convince the indigenous people to cooperate with Britain and France, with the promise that the lands would be given back to the Arabs upon Allied victory. Exceptional circumstances intervened. At the end of the war, however, the Sykes-Pikot Agreement laid out certain boundaries—and the land was so divided that they became British or French mandates. Britain got the Anatolian plateau, Palestine, and Iraq; the French got Syria, and Lebanon. It is no wonder that resentment and anger mark the consciousness of the native people, a situation that needs to be reconciled. The opportunity will present in many ways as the transit is experienced.

Figure 30. Areas of Palestine under Sykes-Pikot

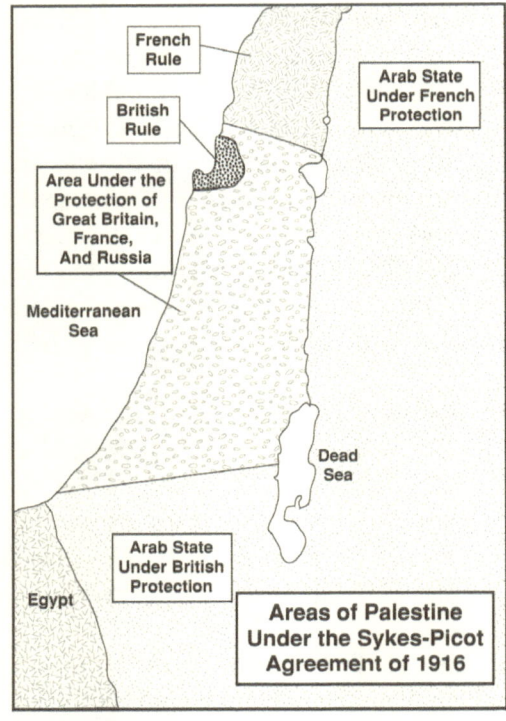

Tribes with Flags, by Glass, provides an interesting and detailed analysis of the social and political background for these divisions. What is interesting here and what complicates the situation is the fact that less than a year after the Sykes-Pikot agreement, Lord Balfour came out for a homeland for the Jews, and the cause for a Jewish homeland had found a champion. Lord Balfour became the strongest, most influential advocate for Zionism at that time, arguing that Palestine might be a homeland for the

Jews. Immigration from abroad began, and quickly accelerated. At the same time, someone muttered, "She is a beautiful bride but already taken." Who that person was remains unknown, but it was a statement that would ring true as history played out in the region. Balfour's Declaration became a standard of the Zionist ideals. A portion of that letter reads as follows.

> His Majesty's Government view with favour the establishment in Palestine of a national home for the Jewish people, and will use their best endeavours to facilitate the achievement of this object, it being clearly understood that nothing shall be done which may prejudice the civil and religious rights of existing non-Jewish communities in Palestine, or the rights and political status enjoyed by Jews in any other country. November 2, 1917.

Reassurances were given that the indigenous peoples would not be threatened, a meaningless promise because it could not be enforced. In this way a second major flashpoint in Palestine came into being—a flashpoint that would later become the State of Israel.

After the Balfour Declaration, a heavy Jewish influx poured into Palestine. World War II only exacerbated the intense desire on the part of many Jews to return to what they saw as their true homeland. After the holocaust, the yearning was deep for a nation where their traditional culture could grow and develop. Spawned in terrorism, indignation, suffering, and determination, the vision of the modern state of Israel became a reality with the United Nations Resolution 181 on November 29, 1947.

Country of Israel

Another challenge arose in the area and by the end of the 20th century, certain events led the United States to commit to restoring the Emirates of Kuwait, one of those clusters of people that, like so many other nations, had previously been under the Ottomans. Iraq, tempted by rich oil reserves, felt justified in moving into the little country. Iraq owed Kuwait a huge debt and could not pay what it owed. If Iraq could conquer Kuwait, and after all the boundaries were not really borders to their way of thinking, they could pick up the rich oil reserves. A Coalition led by the United States reacted when Iraq invaded, and the Gulf War began. This intervention restored Kuwait's sovereignty.

Debt or the threat of it influenced Sadam Hussein's impulse to move against Kuwait. It is obvious by now that the theme of debt is a recurrent one. It pervades society, national goals and interests, and to a great extent the future of most nations. In the instance cited, debt once again triggered violence.

During the War, the public watched on television with complicated anxiety as cruise missiles thundered and crashed into Baghdad. Pictures of the remains of street markets, crumbled adobe houses, and opulent palaces held our attention. Iraq, the mother of civilization, appeared old and worn out. Even those areas of the city that had not been bombed were seen to be in various states of disrepair, a condition in sharp contrast to the gleaming marble of magnificent palace buildings. Something more than restoration was in order. It was Freud who pointed out that a sensation of something uncanny occurs in civilized people when they are suddenly surprised by a home truth they have repressed—a primal fear or desire. Watching the nightly news brought that kind of visceral shock of the forbidden, the hideous, bizarre, sad, yet militarily successful bombardment of Baghdad. It was more than the stony rubble and wounded or dead bodies flung out at unexpected angles in the street that unsettled the unconscious. Our shock related not only to the sheer violence but also to the raw earthiness of ancient memories. The soul of Iraq had been wounded, her beauty spoiled. She appeared hurt, a primal chaos of fire, gray sand, and piles of ammunition. And yet, the striking thing, the decisive thing about the landscape was her resilience.

To the South lay the country of Israel.

Meanwhile, the Gulf War only served to punctuate the animosity between Iraq and Israel, a situation that goes far back in history. In ancient times, King Nebuchadnezzar had captured, enslaved, and carried the conquered Jews to Babylonia. That was in 586 BCE.

Figure 31. Forced Resettlement of Judah to Babylon

At that time, Babylon was the largest and most magnificent city in the world. The Hanging Gardens became known as one of the Seven Wonders of the World. The name of the city was synonymous with the entire empire.

King Nebuchadnezzar goes down in history as having defeated the southern kingdom of Judah which today we know as Israel. The northern kingdom of Judah had been defeated much earlier by the Assyrians. Once again in 721 BCE, the Jews in the north had been defeated and taken away into permanent exile. The Bible records the story as the Lost Ten Tribes of Israel.

There are some today who see parallels to those ancient times. Consider the prophecy from Timothy:

> But realize this, that in the last days difficult times
> will come. For men will be lovers of self, lovers of
> money, boastful, arrogant, revilers disobedient to
> parents, ungrateful, unholy, unloving, irreconcilable,
> malicious gossips, without self-control, brutal, haters
> of good, treacherous, reckless, conceited, lovers of
> pleasure rather than lovers of God, holding to a
> form of godliness, although they have denied its
> power; Avoid such men as these. For among them
> are those who enter into households and captivate
> weak women weighed down with sins, led on my
> various impulses, always learning and never able to
> come to the knowledge of the truth.
> (2 Timothy 3: 1-7)

Students of the Bible make it clear that great military power, vast
wealth, and strength are not enough to protect a society. Destruction
can come from within.

The astrology of Babylon and of Judah was steeped in an
esoteric interpretation of the Laws of the Cosmos. The idea of Sacred
Geometry includes much of what can be seen as the working out of
Divine Principles in our daily affairs. Solomon's Temple was a structure
that incorporated those principles in its actual design. No doubt,
the crusaders, when they encountered these symbols were deeply
impressed with the magic and mystery depicted. It is said that they or
more specifically the Knights Templers, brought back many secrets
to Europe after the wars. These included certain principles of design,
hitherto unknown. When the kingdom of Judah prevailed in ancient
times, a temple had been dedicated to Yahweh, who was the God of
Israel. It housed the Ark of the Covenant. As such it is estimated that
the Temple stood for over 400 years.

To the Judeans, Mayans, and Babylonians Venus was not
considered a benefic power. Her appearance coincided with war,
famine, and destruction.

Some of the Babylonian Planetary omens discovered include:

- If Venus reaches the Sun and enters in to the Sun: a city will be torn down.
- If Venus enters into the Sun and does not come out: devastation, variant: enemy incursion will be in the land.
- If Venus in the morning stands toward the front of the Sun: the land will revolt, there will be much famine.
  Babylonian Planetary Omen Part Three page 45.

This negative connotation is a startling contrast to the benefic interpretation with which Western astrologers honor the "bright star." There had to have been many historical events over ancient centuries that led to this negative interpretation of Venus from Babylon. Ishtar is associated with the Divine Prostitute. Perhaps we can see a parallel to Freud's notion of the id, that instinctual, dark side of us that drives us unconsciously. One can only imagine a long list of ominous events that once recorded were associated with the planet, Venus.

When the Babylonians conquered what remained of ancient Israel, they destroyed Jerusalem and burned Solomon's Temple. This structure and its destruction serve as a symbolic point of convergence. The Temple of Jerusalem had been of great interest to the Crusaders, particularly the Knights Templar, who sought out esoteric mysteries in the structure. The Gothic Cathedrals owe much to the cross-over technology that was gained from a study of the Temple, as well as notable Islamic architecture. Both Jewish and Christian prophecies mention the Temple as a marker point of interest. Fundamentalist Christians relate it to the Apocalypse and the Rapture, a time when "saved' souls will ascend to heaven. These discoveries of geometry and mathematics led to the knowledge that eventually built the great cathedrals of Europe. In this regard, there is a deep connection with Venus, not only because of the beauty that resulted, but because of the mathematical acuity that was necessary to construct the great edifices that dot the map of Europe, even today. Venus relates to analytical thinking and precision.

In any event, old wounds opened and the game of politics and war in the Middle East once again came into play. Later, down the ages, the United States would tie Iraq to the vicious terrorist attack of 9/11.

In March of 2003 a Second Gulf War was waged. This time the United States went into the campaign against the ancient nation without the approval of the United Nations. The stated goals? To locate and destroy all weapons of mass destruction, and remove Saddam Hussein. The vision? The democratization of Iraq!

It is hoped that a fresh and improved version of Iraq might serve as an example toward a new kind of civilization, the next turn in the spiral of history—this in the very location where civilization had first begun, this in the country that celebrates the beautiful Ishtar gates, the gates of Venus.

2010—four years after the 2004 Transit and two years before the 2012 Transit—ushered in what has been termed the Arab Spring. This timing indicates that the Arab World and its desire and aspiration to live by deeper values will be a major theme for decades to come. While protestors marched and flags waved, cell phones buzzed and changed the dynamics of power.

Egypt and Libya are two spectacular examples of how Middle Eastern countries that have thrown off vestiges from the past for a new and better tomorrow. Hosni Mubarak has been overthrown by a people's revolution in Egypt. Perhaps, in even more spectacular terms—because the West came in to assist—Moamar Kadafi was deposed from Libya. The West assisted with technical support on the ground and air support. Cruise missiles were launched. Clearly there was agreement that the dictator had to go and he did. Martian energy had been exercised in what hopefully will have a beneficent outcome with Venus.

Ideally, these transformations and revolutions will provide more humanistic value oriented models for all Middle Eastern countries. It can be said that this kind of dramatic purification was necessary to destroy the ancient repressive thought forms that held the area in a claustrophobic cocoon. (Similar phenomena occurred during WWII with the extensive bombing of Europe and the ultimate destruction of old, restrictive forms.)

The issues in the Middle East are like pieces of chess on the board. The initial game has been launched, at least for our time. These themes will play out over the years up and until the next transit over one hundred years from now.

As we approach the 2012 transit, increasing tension with Iran is on the ascendancy. Ancient wounds, old memories and new desires foment change and conflict. Here, once again we see a county in the Middle East striving for global recognition in contemporary time. Iran, once known as the ancient country of Persia, had fostered ancient civilizations. Her empire reached as far as Macedonia to the North of Greece, to the Indus Valley in the South, and much of the Middle East. Persia had been a leader in literature, philosophy, medicine, astronomy, mathematics, politics and art. Her culture spread far and wide and was greatly admired.

> For thousands of years Persians have been creating beauty. Sixteen centuries before Christ there went from these regions or near it . . . You have been here a kind of watershed of civilization, pouring your blood and thought and art and religion eastward and westward into the world . . . I need not rehearse for you again the achievements of your Achaemenid period. Then for the first time in known history an empire almost as extensive as the United States received an orderly government, a competence of administration, a web of swift communications, a security of movement by men and goods on majestic roads, equaled before our time only by the zenith of Imperial Rome. Will Durant address before the Iran-American Society, 1948

Over the years in the 20th century, a growing dislike of foreign influence and interference eventually culminated in the birth of an Islamic republic of Iran in April of 1979. Today, Iran is a major supplier of oil to a global market, a fact that significantly complicates the issues. Rising tensions with Iran make confrontation with more and more likely as Iran ramps up Uranium refinement in an attempt to join the nuclear weapons club. Add to that Iran's hostile attitude toward the country of Israel, and we see another set of troublesome circumstances that will play out during the ensuing transits.

On the positive side, **Shirin Ebadi** an Iranian lawyer, a former judge and human rights activist and founder of Defenders of Human Rights Center in Iran stands out as a real peace maker. On 10 October 2003, Ebadi was awarded the Nobel Peace Prize for her significant and pioneering efforts for democracy and human rights, especially women's, children's and refugee rights. She was the not only the first

Iranian, but also the first Muslim woman to receive the prize. Ms Ebadi has stated that women in Iran network for peace and justice in a quiet but steady fashion.

> " . . . peace begins from inside: it boils from within, spreads through the family, saturates the society and the covers the international arena. Rather than throwing democracy on a nation through cluster bombs, we must support women and take stronger initiatives to protect their rights." The silence of an oppressed society, whether one that goes through religious or political oppression, resembles the silence of a cemetery." No one can keep a society silent with the threat of a bullet, or at the point of a gun or the punishment of prison."
> Shirin Ebadi, ***2003 Nobel Peace prize winner***

Eight years later, The Nobel Peace Prize 2011 was awarded jointly to three women—Ellen Johnson Sirleaf, Leymah Gbowee and Tawakkol Karman *"for their non-violent struggle for the safety of women and for women's rights to full participation in peace-building work"*. There is a general sense that the work of women is on the ascendancy and this will only gain momentum with the Transits of the 21st century.

There are other flashpoints in the world, but at the time of this writing those mentioned above seem to be the hottest. Terrorism and its ramifications are deeply tied to Asia as well as the Middle-East. Poverty, exploitation, and the added element of fanaticism, only contribute to its growth. This point was emphasized in the discussion of the first transit of Venus in 2004—the transit that was seen mainly in Asia and Eastern Europe. This is not to say that these events are not of a world wide nature. They are. But it is in the East that the transit was initially seen.

There is additional reason for optimism. It has been predicted that the Sixth Ray of Idealism is gradually fading out. The notion of idealism can be tied to fanaticism as is evidenced in religious, racial, and political attitudes that cannot or will not tolerate "the other." Nothing but one's own view of life, God, religion is correct and some are willing to die to prove the point—all this relates to the Sixth Ray. The turning point, 2024 comes soon enough after the Transits, indicating that a major shift toward a more unified planet is a real possibility. The glamour of separatism can then be seen for what it really is—a deceptive glamour—and gradually dispersed.

Rays ebb and flow in their potency in long term cycles and the study of such cycles can yield important clues toward the future. Around the same time, it is predicted that the Fourth Ray of Harmony through conflict will come in at the year 2025. This powerful influence will gain in ascendancy for some time. For the immediate future, it can complement the energies from the Venus Transits.

The energies of the Seven Rays are at play within the fluid fields of human endeavor. Each nation represents a collective life, a unity, with varying qualities and patterns. These are characterized by differentiated energy and resulting structures. Planetary, and even more subtle energies, project themselves in time and space, bringing all things into being upon the plane which they project themselves. Underlying this complex process—a great evolutionary urge.

Is the Venus transit of 2012 a critical ingredient in human affairs? All the factors so far mentioned in the transformation of the Middle East come to bear on the fate of civilization there, but also, and more importantly, on a worldwide scale. Modern globalization, with its accelerated rate of change, has spun itself out in a painful, sometimes brutal, process. Not only have languages disappeared but also peoples, ecosystems, and religions. The illuminating, refining qualities of Venus could be a blessing. Breaking loose from old constraints, finding new and better technologies, uplifting through art and beauty—these are the ways of Venus. On the other hand, the reshuffling process could lead to deepened unease, leading to war and even further destruction. To gain more insight, we will look at the second of the 21st century transits.

Chapter Ten

THE 2012 TRANSIT

A darkness all of light. The more
You tried to hold it back, the more sweetly and irresistibly it arrived.
Franz Wright

Figure 32. Boucher, The Triumph of Venus

Viewing and appreciating a work of art forms a magnetic field around the images and their meaning, thus intensifying its impact. For that reason, Boucher's Triumph of Venus is worth considering, especially as we approach the 2012 transit of Venus. The goddess reclines on a rock surrounded by lovely female divinities, cherubs, and exotic mermen. As the glorious centerpiece in the midst of churning, energetic waves, Venus is clearly at peace and at rest, even though darks clouds and deep green waters swirl around her and her companions. To the far right, above her on a hill, there is a hint of an edifice—the form suggests a temple. But Venus has come down from the heights to the sea, where she is clearly in command. The cherubs unfurl a luscious banner of rosy, satin-like material over her head. They frolic in the misty air, tossing the banner. The mood is full of joy and play, despite the dark color of the water and rocks. Venus and her troop irradiate the whole space (Earth) as she takes her rightful place within a wider network of activities. The goddess has the ability to bring the mind to a new zenith of creative power.

So many indicators converge in 2012—the end of the Mayan calendar, the Hopi prediction, biblical prophecy, the three and seven year cycles for group work, the number 12 as a number of completion—that we have to take very seriously the additional factor of the Venus transit. The planet Venus, as a palpable presence in the formation of cultures, will have a definite impact. There is a natural tendency to be enthralled by her power to bring light. Her yearning for love, as expressed in captivating goddess images down through the ages, exemplifies the desire to be seen—to have the innermost beauty revealed. Such a desire brings with it certain risks. It invites exploration into the terrain of eroticism, gender, wealth, and extravagance. Looking at Venus as she transits the sun, we discover sunlight.

We have looked at the maelstrom of chaos, competition, and conflicting goals in the Middle East as a major factor in shaping and reshaping the world stage. The future rests on how we mange not only the social issues but also the complex tools of war and weapons of mass destruction. These weapons no longer fall into the sole domain of Vulcan, with his skills in fashioning armor and metallic weapons. Uranus has entered the picture with the potential for nuclear destruction. Then there is the threat of biological warfare—a unique combination of Chiron, Pluto, Mars, and Venus.

This treatise rests on the concept of a spiritual world, with visible counterparts in the sky that run parallel to our manifested life. From the vantage point of astrology, we bring some powerful tools into play as we build our own space. Events, both personal and collective, reverberate from the past as we strive to generate superior wealth and/or power and well-being, to create a more advanced culture.

The second in the pair of twentieth century transits takes place on June 6, 2012. It will be visible from northwestern North America, Hawaii, the western Pacific, northern Asia, Japan, Korea, eastern China, the Philippines, eastern Australia, and New Zealand.

As astrologers we seek to find certain parallels in consciousness, both local and universal. We seek to determine ways in which humanity may be more and more able to express the Divine Immanent, the Divine Transcendent, and to register the illumined awareness that Venus can bring. Once again the transit takes place in the sign of Gemini.

2012 Transit of Venus

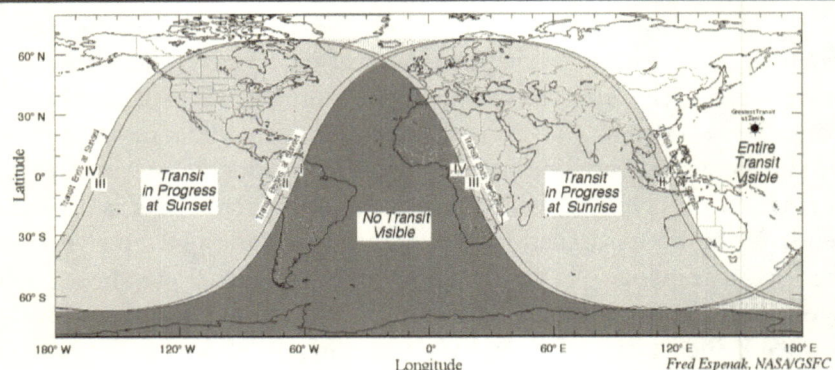

Figure 3 - World Visibility of the Transit of Venus — 2012 June 06

Figure 33. Venus Transit 2012, Visibility at Sunset

Venus squares Mars in this transit. Venus has to do with interior beauty and happiness. She also rules adaptation and all forms of mutuality. Mars tends to bring in a harsh component—an aspect of quarrels, if not of war. Conflict of some kind is assured with the Fifth Ray opposed to the Sixth. If the latter is transformed from fanatical devotion to true aspiration, an opportunity that was delivered at the 2004 transit, this could portend real progress for humanity. In this sense, conflict can work out on the mental plane and not necessarily explode into war and physical violence.

The first transit was likened to a great rehearsal, placing humanity in a position to choose or resist the prospect of a more peaceful world. The 2012 transit indicates the culmination of that critical period. Danger is accentuated by the fact that the Sun squares Mars. The red planet, with its male point of view, can be taken in three ways: the eyes of love, eyes with desire for mastery, and the gaze of honor and enjoyment. The key here is that desire is linked to beauty. A battle between selfishness and the greater good is promised. It need not end in disaster, but things will be hammered out. Vulcan will be in play. There is a tendency to run risks with this aspect, leading to the precipitation of combative, opinionated expression. This is the kind of aspect that led to the crusades. It is an influence that causes participants to want to fight something—the tremendous energy that comes with the Mars square aspect brings a restlessness of spirit that does not easily give up.

With the lessons being learned from Pluto in Capricorn, we understand how upheaval could be exacerbated by crises in terms of national identity as well as environmental or natural disasters. Prevailing conditions can be even more aggravated by Mars, the god of War, squaring Venus, the harbinger of peace and beauty. There is no doubt that these configurations will bring crisis—a time of great difficulty, and/or a turning point. A crisis can be a great transition. For those who linger too long and cling to the past, the lesson put forth in 2004 could be lost.

Crisis emerges when the need or the imbalance is great. What is important for this discussion is that the Venus transit of 2012 brings into play the experience of eight years of evaluation and recapitulation going back to the 2004 transit. We might call this the inciting incident. The two transits, taken together deals with the kind of paradigm change that challenges the collective will, stimulating desire and the imagination to create something new. Intense suffering is suggested if we hold on to the dying past. There is no doubt that a new threshold is being presented to humanity. Not to cross it will lead only to other more drastic thresholds laden with even more suffering and pain.

Intuition and imagination are necessary to create the pathway between one state of consciousness and another. In this regard, the Venus/Mercury conjunction can prove invaluable in terms of being able to see the crisis in a new context. Mercury represents Hermes, and Venus, Aphrodite. Together their conjunction forms the ideal of the Hermaphrodite. In terms of the psycho-physical plane, a more androgynous way of thinking and communicating is in order. This is the way of Aquarius.

Sometimes Mercury has been known as the three headed one—Sun, Venus, Mercury. Here they are so closely conjoined during the transit that it can be said that they act as a single unit of light. This illumination can provide real impact if brotherhood and a new vision can overcome the real intensity of the Mars square. This can be a period of brilliant thought. Mercury, on the higher level, points the way to higher states of consciousness. With the intensity of the Sun and Venus, in this case, revelation of next steps and changes required can certainly occur.

Venus trines Saturn—this sets the stage for a positive vehicle of planetary soul expression. Venus refines the nature of the collective with respect to our responsibilities to others. Right human relations is one of the goals of the New Age. The Venus/Saturn connection can be helpful in structuring new patterns in society as vehicles of good-will. The potential is there.

Sun and Venus square Neptune. There may be a craving for something great, but wisdom must be sought out in order to bring it about. Ideals can be carried to injudicious extremes, for the tendency of Neptune to produce scandal and glamour is always real. In practical affairs, it will be wise to guard against deception. Mercury also trines Saturn bringing a potent combination of mental energies. Mercury carries the Fourth Ray, and thus there is the potential to create right relationships for the expression of Mind—Venus. Saturn brings the Third Ray, and the ability to work on those deep structures where ideas can take on power and come into manifestation. Streams of new (young) ideas result in the conscious structuring of interactive relationships. There must be a clear vision of the desired outcome, as well as the ingredients for that vision to be available.

As we saw in the myth, Venus and Mars were destined to be lovers. The gods rejoiced in their union—a union which, of course, exemplifies the highest expression of both energies of Venus and Mars—a major challenge for this period. Will it be love or will it be war? Perhaps it will be synthesis, and there are stunning clues that this may be in the offing.

At first we may not think of precious stones when we consider the transit. It is important to recall, however, that the diamond is associated with Venus—it signifies the crystallized focus necessary to open the third eye. The diamond is 'her" stone.'

Synthetic and cheap diamonds are now a reality. Science has found a way to refine the element of carbon into a beautiful translucent stone. The diamond has long been a symbol of great beauty and only more recently a symbol of enduring love. It would appear that the magic of science and technology has made more light available—it will be concentrated and effective.

> D.K., the Tibetan, tells us that the Divine plan is "hidden in the geometry of a crystal, and God's radiant beauty stored in the color of a "precious stone." Further we are reminded that, the truly illumined man and all who have taken the three highest initiations are always referred to as "the diamond souled." They, in their totality, constitute the 'jewel in the lotus'—that twelve-petalled lotus which is the symbol and expression of the potency of the planetary Logos."
>
> The diamond is the model for the gem our own consciousness can become, and is now physically within our creative power to quickly synthesize. This new development, if we carry the symbol forward, suggests transformation in the new age can be exponentially accelerated. *Thoughtline* November 2003, Miki Webb.

It has been said that in order to bring about a Synthesis of planetary magnitude, a whole new order of wisdom devas must come into the world. These are the little builders who have synthesized the 1st, 2nd, and 7th Rays.

Figure 34. Awakend Earth Deva

The devas operate on a parallel path to our own. Perhaps the creation of synthetic diamonds is just one omen—a foreshadowing of future works of art and more widespread synthesis. Perhaps the new order of devas will contribute to the revelation of our own inner jewels. This second transit is especially meaningful for the United States.

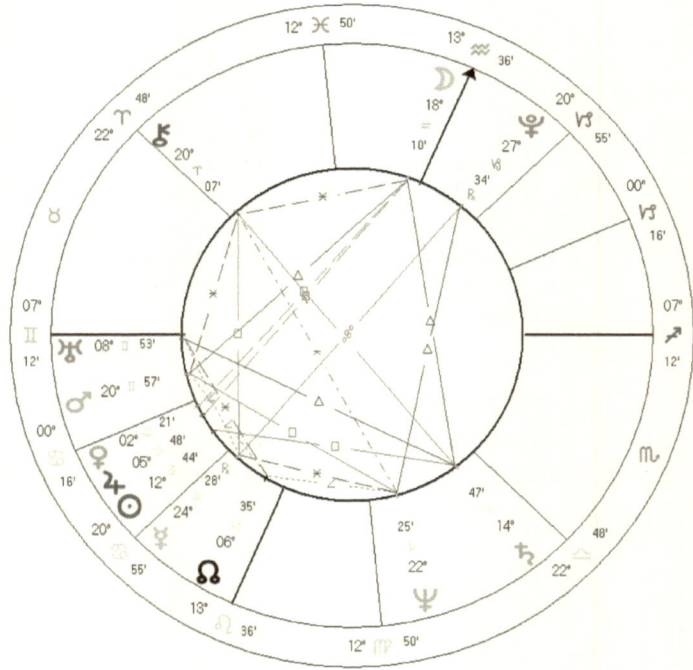

United States
Event Chart (2)
Jul 4 1776 NS
2:13:32 am EST +5:00
Philadelphia, PA
39°N57'08" 075°W09'51"
Geocentric
Tropical
Placidus
True Node
Rating: A

Figure 35. U.S. Chart

The second Venus transit of the Twenty First Century falls in the Second House in the United States chart (with the Gemini Ascendant.) The Second House relates to issues dealing with money, resources and finance. From the esoteric angle it can signify how spiritual life is expended. It can also refer to an inflow of spiritual values. All three planets—Sun, Venus, and Mercury are in close proximity. Venus and Mercury surround the natal Mars like bookends—a very useful aspect, carrying with it the potential of harmony linked to clear vision and excellent communication. Mars brings courage and energy. These are good aspects for diplomacy and for resolving opposing energies. One could say that there will be an intensified interest in Cultural Diplomacy which will gain in power as a central part of national strategy. Often this approach is termed Soft Power which relies on the ability to shape the preferences of others through co-option and attraction. It sharply contrasts to "hard power" which is the use of coercion and payment. Soft power is the ability to shape the preferences of others and get others to want the outcomes you want. Soft power resonates with Venus.

This kind of focus will include the support of cultural activities abroad, funding intellectual publications, and a revitalized use of the U.S. Information Agency. The Second House can be seen as a storehouse of life/soul energy.

The conflict between the Fourth, Fifth, and Sixth Rays—Mercury, Venus, and Mars—works out the conflict between the Fourth, Mars, and the Fifth, Venus. Human relations, the arts, and professional spheres will be influenced. Transiting Mars can enliven the Fifth House of Creativity where Neptune naturally resides in this chart. Transiting Mercury squares the natal Neptune, an aspect which can enliven the artistic and aesthetic arena and may give a spiritual or fantastic spin to its expression. There are elements of self-sacrifice, but it can promote increased spiritual insights and perceptions. Neptune rules spiritual desire, or the urge to manifest the Christ. And with Chiron on the Midheaven there is the possibility of wounding and/ or the infusion of Christ energy. Past history relative to resources can be an issue with Pluto and Mars over the Eighth House. This can also stimulate transformation at a very deep level. Just as Europe held a supreme position in the world during the 19th century transits, the United States reached that pinnacle during the 21st century transit. Now, there is an opportunity for the United States to play a major role in establishing freedom and liberty in many parts of the world and to become a cooperative global player shaping events with a Venusian twist. (Spiritual values, life values, a higher morality.) The closest aspect: transiting Sun at 16 degrees Gemini sextiles the 12 degree Cancer Sun of the United States chart. Gemini is a mutable sign. This influence suggests a gathering in of experience. Yes, there will be constant flux, but this is a marker event in the passage to a New Age. The Second Ray appears very strongly in the idea of relatedness and connection. After all, Venus, the Esoteric Ruler of Gemini, seeks to transform all the issues of duality through the creative expression of the mind with respect to the soul. This transit will usher in a period of constant change, as the United States persona reorganizes in terms of soul expression. And soul expression is really the Christ principle, the principle of Love. Gemini influence can change, mutate, and adjust whatever is necessary for the growth of the soul. Since Gemini also carries the energy of Sirius (another great Second Ray vehicle in our solar system) the combined Sirian/Gemini effect will bring a huge dose of Second Ray influence. Love/Wisdom will flood the planet.

Mercury is in the sign of Cancer, a most interesting position. Here it can bring in those energies which bring about emotional security and thus it resonates deeply with Venus. The Cancerian connection brings Neptunian energies into play and with it the 6 th Ray. On a higher turn of the spiral this can stimulate devotion to the Plan. The United States will have access to energy which will encourage greater integration of ideas to sustain and nurture humanity. This is a real potential.

Jupiter trines natal Pluto. We have been told that Pluto and Venus make things happen. Jupiter, in this case at 28 degrees Taurus, trines Pluto at 27 degrees Capricorn—an opportunity to manifest needed resources. Vulcan's influence may be at work, as we learn collectively that there is no difference between the material and the spiritual spheres. This powerful combination of Jupiter and Pluto can destroy obstacles as it brings in both the First and the Second Rays—a transformation and expansion of consciousness is in order. It is a positive aspect for spiritual regeneration.

Let us remember that Venus represents the urge to merge. Within the maelstrom of events that have occurred and will occur during the few years of the transits, Venus is the magnetic force that can bring diverse people and conflicting ideas together. Venus harmonizes.

She can influence deep seated tendencies toward beauty and balance. A similar inner experience can be felt on the inner, or soul level. In other words, a greater coherency of being is possible not only on the group but the individual level.

Such a metamorphosis cooperates with natural law. This kind of transformation displays movement on many fronts, as it seeks to synthesize past learning and accomplishment. The chart incorporates a profound world of change. What happens depends on how the United States, and Humanity in general, respond to the infusion of energy and the degree to which the Principle of Goodwill becomes tangible.

> Your holy mind establishes everything that happens to you. Every response you make to everything you perceive is up to you, because your mind determines your perception of it. A *Course in Miracles*.

Chapter Eleven

Paradoxical Lightness

Conclusion

The universe, and our planet within it, is a series of creative transformations, each with its own flow and quality of energy. The Venus transit, though only short in duration, deserves to be examined because it suggests a social cycle of change and transformation. As individuals, we are affected by the cultures in which we live, and to that extent the transit will have a socio-personal effect on us. The Venus Cycle last occurred 129.5 years ago, in 1874, and again eight years later in 1882—a pattern of 121.5 plus 8 years. The 19th century cycle is of particular interest to us as it serves as a kind of benchmark from which we can extract meaning. Then moving forward, in terms of the 2004 and 2012 transits, they suggest the making of still yet another transformation on Earth—a transformation that requires illumination to take us out of our present destructive and inefficient modes into one that will renew the planet and all her inhabitants.

The light of Venus presents a paradox. She is the patron of civilization, and yet in order to survive, the civilization we have created needs a major overhaul. Venus keeps the right company to accomplish the task. Her husband Vulcan is the master artisan and craftsman. He can inspire the technology that will meet the need of a new age. Pluto (with whom Venus spent some trying days in the ancient myth) brings light out of the depths and out of the darkness. With the convergence of all three of these energies—Pluto, Venus, and Vulcan—life unfolds in new forms.

"This capacity is indestructible and is itself a divine focus of energy which must and will without fail carry forward the good undertaken under the inspiration of The Great Architect of the Universe. He fashions all things to a divinely foreseen end and in this sign—through his agents, Venus and Vulcan, typifying the form and the soul—will lead man from the unreal to the real." *Esoteric Astrology*, page 404.

Perhaps Venus is telling us that the creation of the next civilization requires us to cooperate more fully and more consciously with the unseen energies. This requires a participation in the beauty inherent in the moment. Our thoughts, both positive and negative, are the builders of the future. This is where art, music, and popular film can have an enormous influence toward creation or destruction. There is no doubt that the 2004 transit is positively influencing the arts. Beauty, with her illuminating power, can shed light on the obstacles between us and our Source, between us and humanity. It is as if Venus were saying, "The door to beauty is right here in front of you. Open your eyes, step up and into the moment and see. Be yourself."

The eye to be opened is of course the third eye, the eye of intuition. Opening the third eye brings that kind of seeing that knows the truth directly, pierces to the heart of it, and radiates and draws to itself magnetically. This is the function of Venus, the illuminator, the awakener.

Beauty's Divine Light will facilitate the permeation of the particles of manifestation with Light. The sense of separateness fades in this light, and the discovery of the path opens to awareness.

In terms of world development, Venus is important in the Gemini Festival of the New World Religion as some sort of synthesis in the area of religion unfolds.

> "In the coming world religion this fact will be noted and in the month of June, which is essentially the month in which the influences of Gemini are particularly strong, due advantage will be taken in order to bring man nearer to the spiritual realities. Just as Venus was potent in producing the relation of such pairs of opposites as the fifth kingdom of souls and the third kingdom (the synthesis of the sub-human kingdoms) leading to a Great Approach between soul and form, so in the new world religion this fact will be recognized. Appeal will be made to the Forces which can utilize this planetary potency in order to work out the divine plan upon the Earth"
> *Esoteric Astrology* page 355.

If the plan is to work out on Earth, it must take form. It is a mistake to think that Venus works alone in this gigantic task. That is not the case. If we take the Moon to represent form, then Venus, in concert with a number of archetypal females who figure in several of the constellations—Cassiopeia, Coma Beatrice, Andromeda, Virgo, and the Pleiades—provide insight into the progressive unfolding of form.

Venus has a transformational role to play in the new age. Her influence will draw more realized souls to the planet. It has been said that Venus with Pluto are dominant in bringing about world results. Astrology literally means seeing the logos in the stars. The Divine Teacher works through all creation to share mysteries and motions that hold the keys to our evolution. Deep energy is invoked with the transit of Venus. We are irresistibly drawn to her beauty, and thus our intelligence is awakened and the sacred is activated.

The energy of humanity is shaped by archetypes. It is hoped that this compendium and analysis is helpful in such a way as to play a small part in the stimulation of critical thinking and the comprehension of new and better designs. During the period of the Transits of Venus, we are called upon to make serious adjustments in values and paradigms leading to the intelligent reshaping of things. In this process, Venus stimulates the drawing out of beauty and language is unfolded.

We live in language, and today we live in the language of the internet. It is there that beauty, information, and knowledge collide—

the output is there for all to use. Mercury forms a stellium with the Sun and Venus in 2012. In ancient times, Mercury was often considered the Sun because of their close proximity.

Mercury and Venus can never be more than 72 degrees apart. There is always potential harmony with this aspect since Mercury can serve to resolve opposites. Venus can harmonize with her spiritual and soul qualities.

The three planets in Gemini strengthen Mercury as a herald of the New Age, with all the possibilities of networking that are implied. The advancement of electronic media can facilitate this networking aspect, and already new schools, such as Christ Church College of Trans-Himalyan Wisdom in New Zealand, The Dragons of Wisdom in Northern Europe and Morya Federation and Morya College. The latter is an online place of study. Students from all over the globe can participate in meaningful way to learn about Esoteric Psychology, Esoteric Astrology, and the Deva Kingdom in a great quest to further knowledge and gain wisdom.

Mercury has a close tie with the United States chart as the ruler of the Ascendant. The greatest accumulation of wealth still resides in the United States. Economics remains a major driver in the affairs of mankind; China holds a huge proportion of humanity within her borders. Will the U.S.A. and China develop the mental know-how to come to terms with technological advancements? There is hope. Today there are a large number of people who once played out the Atlantean drama. They carry the experience and the will to resolve the problems of humanity, so that we can move on to a future in harmony with the light of Venus. The conflict will revolve around the use of the magnetism of Venus on a global level. Mercury will be helpful. It is involved in media and commerce through communication. It is also the sign of youth. Gemini, as ruler of the United States Ascendant, can vitalize through dialogue and other forms of communication, and help bring our Chinese partners fully into the New Age.

> Recalling that the Sabian symbol for the 2004 transit that initiated the activity. "Two Chinese Men converse in Their Native Tongue in an American City." Chinese involvement will be significant. The keynote: "The need for the mind to retain its independence from its physical environment in order to concentrate on its special problems." *Astrologcial Mandala*. Page 101.

China has entered the space age, and the United States has backed off somewhat with the end of the Shuttle Program. China can and will become an enormously important influence in the world marketplace, and so it is that interaction between China and the United States will produce enormous transformations. Let us hope that they will be fruitful and in harmony with the best interests of the Earth—the potential is there. China's culture is far older than that of the United States, and carries the memories of Confucius, the Tao, and Buddhism. The interaction between these two great countries can lead to a new realm of activity that will be strange and yet promising for both. Caution must be taken by both counties, as they struggle to meet their voracious needs for resources of all kinds. China has ascended as a world economic superpower, second only to the United States in its contribution to the world economy.

There is a trend toward political liberalization that can come to fruition during the period of the transits. Opportunities for cultural exchange will flourish, and China will embrace the information society with open arms. Why not? Soon China will be a major supplier of both hardware and software products. The United States and China, in their mutual search for economic opportunities, will transform China's role in the world market place.

The downside of China's emerging role in the global market and that of all Asia is the growing danger of AIDS. By the last of the 21st century transits, Asia will have had to deal with an epidemic that can be clearly likened to the Black Plague of the 12th century. There are ominous predictions from the United Nations that Asia could follow in Africa's wake where in some areas the infection rate has soared to 39%. Pressure from this epidemic can lead to further international cooperation.

What is stressed in the keynote of the Sabian symbol is the influence of the mind. Venus has an important role to play if we take her light to be the way that true mental independence can be achieved. She is the link to the eternal and to the transcendental field of awareness. Creative Intelligence is needed to live in the world and yet be not of it.

The Sabian symbol for the Venus position of the 2012 transit: "A woman activist in an emotional speech dramatizing her cause."

The keynote: "A passionate response to a deeply felt new experience." *Astological Mandala*, page 100.

Women will come into ascendancy as never before. After cycles of terrorism and war, the feminine will revitalize a tired humanity. If at the very least, nothing is done but the restoration of Beauty, that would be enough to have a world-shaking effect. There is hope for a culture that relies less on militarism and more on one that seeks peace, prosperity, and security through respectful negotiation, education, and development. Already women have stepped up on a global level. As an example of this contribution is the Women Peacemaker program at the Joan B. Kroc Institute for Peace and Justice located on the campus in San Diego, California. The program pairs each participant with a Peace Writer and a documentary film team to document her story. Lessons learned include approaches to peacemaking and justice thus increasing the capacity to participate in conflict resolution and peacebuilding efforts in post-conflict scenarios. The stories are posted on the internet for all to see and all to use.

Building on what women have learned in the field, there is the possibility of bringing experienced women peacemakers together. They can cooperate by sharing stories and experiences and thus leverage the past into a more just and peaceful future. Efforts of this kind are already under way with the cooperation of the United Nations.

U.S. connections to Asia will be vitalized, especially with China, as we seek to create a synthesis of what they bring forward from their ancient past. It must be remembered that the Chinese can trace their heritage back to ancient Atlantis and we, the newer Aryan race who inhabit a country with a Gemini ascendant, have the vitality and fresh point of view that demands a different kind of leadership. But to do that, Lady Liberty, as symbolized in the Statue of Liberty, will have to take a fresh look at the American pursuit of utopian urges. With China as a partner, we can establish a long term dynamic potential based on newfound insights.

The need for communication is explicit in the symbol. To be effective, the eternal feminine must be brought into play. The issues need to be emotionalized in such a way that the masses can be receptive to new and innovative ideas. Here we can see the 6th Ray come into play with mental energy that is passionately moved. The downside is that passion can turn into fanaticism, and the attempt to gain the field can give sway to violent means.

Within the 21st century transits of Venus converge seed themes, problems, and issues that will bring about profound cultural transformations later in the century.

Some of these include:

Genetic Engineering will transform health care in the 21st century. (At first the process will be retardation, but later, as the species becomes stronger, major diseases can be eliminated.)

Accelerated use of artificial intelligence—first in the developed nations and then globally. Mercury and Uranus will be highly active as the New Age settles in.

Widely used instant messaging, cell phones, internet—significantly impacting the political arena.

Further inventions will encourage unity via communications.

Magnets will be used as a source of power

Non-polluting sources of energy will be available and affordable once research and development underway during the transit period yield positive results. This effect will be felt early in the 21st century. Multi-fueled options in automobiles will become common.

These are just a few of the kind of changes we can anticipate. The most important will be a growing sense of unity among the people of the earth—a kind of integration that is imaged in the movement of the planet.

Tracing the pattern of Venus' movement, in a little less than eight years Earth time and thirteen years Venus time the following pattern emerges. What a beautiful signal from our sister planet—a cosmic star.

Figure 36. The Pentagram

Figure 37. Five-Pointed Star

From our vantage point we watch the heavens. Over time, Venus makes the pattern of the pentagram in the sky. She unfolds a secret as she draws us from intellect to intuition. The pentagram is the five-pointed star. One arm, representing spirit, points up—the other four points represent the four elements. The pentagram is the simplest of stars, and has long been seen as a symbol of protection. During the Middle Ages, the five points of the star represented the five wounds of Christ. As far back as the Mesopotamian culture, this symbol was incorporated into folklore and myth, and has been found at the sites of ancient temples. The number five has long been associated with the incarnate human being—we have five senses, five fingers, and five toes.

Our very constitution holds the promise of ascension. Michelangelo's human was depicted in terms of a human star with five points, head, arms, and legs—an example of symmetry. It can also indicate generative and destructive cycles as well as indicate direction. It is a symbol for Venus, and for the 5th Ray of Mind.

With the 21st century transits, the role of women will be transformed and elevated to new heights. A look to the past to justifies this assertion. According to a website, the history of the 16th century transits of Venus provides a great and notable example—a period when women assumed positions of immense power and prestige. This series of events presents a milestone not only for the past but also one that points to the future. Moreover, the 16th century meteoric rise of women to the world stage can and should be taken as a precedent. While it can be said that Venus had not yet been discovered by European astronomers, this certainly did not diminish her cosmic effect. Her energy was available then, just as it will be for the immediate future.

After the 16th century transit was experienced, leadership fell to women in ways so new and broad that we must, in retrospect, admit that Venus was involved. Critical analysis, innovation, and the capacity to create new thought forms prevailed. What must be emphasized is that then, as now, women were definitely on the ascendancy—Elizabeth I, Mary Queen of Scotland, and Catherine de Medici all anointed Queens—are included in the list of prominent women of authority and genius.

Simultaneously, the Protestant revolution offered women new, dramatic and varied opportunities—many joined the new religion and found paths to different and often improved conditions. The Protestant movement had shattered old barriers and opened new doors for many. At the same time Catholic women made their mark as counter revolutionaries. The history of the time provides stories of great female mystics, devout women all over Europe who sought to imitate Christ through extreme acts of self mortification and intense devotion. Their dedication and sacrifices often manifested as mystical experiences and rare revelations. Some of their experiences have been documented such as the life of the great St. Theresa of Avila. Two years after she was born, Luther had started the Protestant Reformation. Within the context of change, Theresa stands out as one who, in counter point to the "revolution," sought inner peace from outer turmoil. Like Galileo, her now famous writings challenged the inquisition. Theresa had relied on inner guidance and clear thinking—her message was unique and had to be considered deviant from the prevailing authority of the church. In the end, after much persecution and harassment, and many years later, she was made a doctor of the Church. This is a rare honor bestowed on those who have been recognized as having been of special importance in terms of theology and church doctrine.

Less than twenty years before Teresa was born in 1515, Columbus had opened up the Western Hemisphere. New waterways as well as improvements in sea craft and ocean technology stimulated European colonization of the New World. Expansion of all kinds— geographical, spiritual, social and mental, marked the transits of the 16th century. Waterways were explored, new connections made and the world was never again the same.

To add to those changing times, prostitution offered still another possibility for women to break from the past and find some modicum of independence. Municipal brothels came into existence. On a different level but still highly visible, upper class courtesans, a much rarer and more privileged group, found patrons to promote their position and also their security. This type of activity with its sexual overtones draws a parallel to ancient Babylon and temple worship since it is well known that sacred prostitution has long been associated with the planet and goddess, Venus/Ishtar. To prove the point, her holy city Uruk was known as the town of sacred courtesans and the goddess herself was reputed to have had many lovers.

Lastly, there is its relationship to the Golden Mean. Each of the tips of the pentagram form perfect isosceles triangles wherein lies the secret of beauty and of proportion. And so it is that Venus winds her way through the heavens leaving patterns of truth behind for us to penetrate.

The 5th Ray has many names including: The Revealer of the The Truth, The Great Connector, The Crystallizer of Forms, The Three Fold Thinker, The Door into the Mind of God, The Angel with the Flaming Sword—to name a few.

The dynamic of the 5th Ray encompasses an unremitting urge to penetrate the veils of form and matter until their secret causes are revealed. The 5th Ray can focus like a laser and seeks to discover the keys to nature mysteries as we saw in the charts of Darwin and Einstein.

In the 21st century transits, Venus trails the sun, which means that the sun is farther ahead in the Zodiac than Venus—a condition which characterizes Venus as Lucifer and the Morning Star. When the Sun trails, Venus has been traditionally associated with Hesperus, the Evening Star, as is the case with the 2012 Venus/Hesperus transit. Dane Rudhyar associates the Venus/Lucifer phenomenon with a new quality of will and purpose. The Venus/Lucifer relationship is more deliberate, with a distillation of experience and an elaborate set of guiding principles. This certainly could be the case if China gets more involved and interactive with the United States in a meaningful and constructive way, thus setting the stage for a new world paradigm. Will China step forward and help ameliorate the world debt crisis? The Asian influence can bring in long-term planning based on established values and long tradition. Will the Arab Spring attain its idealistic goals? Historical precedent will play its part as new values and ideals are integrated—a synthesis of historical precedent plus fresh outlooks.

It is not likely that the potential of the Venus transit will be realized at once. Just as in all the previous transits, it will take time for the potential to be activated. But as we have seen from a little glimpse of history, the transit of Venus cycles can bring strategic stages of social development and the refinement of culture. All the while, humanity remains open to new experiences of Venus, experiences that can be courageous and optimistic. Of course soul awareness is key. Soul implies the inner life, the vertical life—it is what links us to our ultimate source.

If Beauty is rescued for the sake of life, world-shaking empathy will be released, for Beauty is always associated with a furtherance of life. We have seen how this release has played out in past transits in terms of art, government, science, and various other forms of social relations.

We have seen particularly through art how identification with the beautiful, the sensuous, and the joyful can elevate our consciousness and change our self-concept. All the qualities that Venus brings help to create the necessary stages of growth and cyclic development of the spiritual life itself. These beautiful gifts of knowledge and wisdom for all mankind are emerging as Venus makes her procession through the skies. The transits of Venus are but brief cyclic events that release energies of love and of the 5th Ray of Science. Tremendous breakthroughs in science and technology will be made following the

21st century transits. It is likely that the "theory of everything" that Einstein sought will at last be discovered and accepted. The Cosmos affects humanity via the changes of patterns, sun spots, moon cycles, births and deaths of stars, and new configurations in the heavens. But there is a reciprocal arrangement because we, too, affect the Cosmos. Our thoughts, feelings, and actions create beauty or dissonance.

At present we are living in a world where Beauty has been hidden with unalterable and dire results. The eye of Humanity looks up to the Hidden World of Spirit and Inspiration but the vision is blocked by astral clouds that surround us and our planet. Similarly, it is difficult for the higher energies to penetrate downward to humanity because the obscuring clouds hover as a barrier. But despite the problems, Venus is magnetic and draws our attention. We can feel her pull even if we cannot yet clearly see her. The paradigm might look something like an hour glass:

Hidden World
Beauty Veiled
Illusion
Glamour
Maya

Astral

Ordinary Eye

Astral
Power
Sex
Money
Property

Table 8. Desire World

Such a world view, teetering as it is on the threshold of opportunity or dissolution, requires a reordering of priorities, vision, and values. The Libra quality of Venus speaks to choice and the meticulous weighing of ideas. With Saturn as the esoteric ruler of Libra, justice permeates the decision making process—the scales must balance.

On the political side, this transit should energize the passage of the Convention on the Elimination of All Forms of Discrimination against Women (CEDAW) in the United States; it has already been signed by many countries. By the second transit, the ideas embedded in this Convention will have had global attention and resulting awareness. Additionally we in the West need to study and better understand the impact of Sharia law, since at its core there are basic inequalities between Muslims and non-Muslims, and between men and women. Already in Britain, where Sharia law is being used in tribunals, there are conflicts in that it compromises the tradition of equality for all under the law. There must be a way to ameliorate these differences, especially if it threatens the fundamental values that underpin Western society. We can envision a kind of duality where vision upward and downward is blocked. Old forms cling, and crystallization sets in. In the above model, the eye of the Higher World seeks to penetrate into our world of Piscean constraints. Looking downward, higher vision seeks to manifest on the earth plane through service. Simultaneously, the eye of Humanity looks upward to Beauty and Purpose. If it can penetrate through the clouds of the astral plane, it will see afresh, a critical mass is formed, and a transformation can occur.

Hopefully, we will choose Beauty with Love that is progressively expanding as the Seventh Ray of organization sees and subsequently facilitates Spirit in the world even when it not yet physically present. If so, the Desire World is transformed into a Transforming World. A possible design is depicted below.

Spiritual Guidance
Higher Purpose
Beauty
Light
Love
Joy
Third Eye
Transit 2012
Synthesis
Transit 2004
Third Eye
Joy
Sex
Power
Money
Property

Table 9. A Transforming Civilization.

Within this context, the 2004 and 2012 transits of Venus act as bookends of consciousness. The space between the transits provides the time to reformulate the purpose of Humanity and make intelligent adjustments to circumstances. The gifts of Venus will stimulate psyche, and provide a necessary period for psychological growth and further development. A major modification must be the reinterpretation of Islamic culture and law by both East and West.

Proper significance as well as appreciation must be given to the Muslim tradition and culture. It was the great Caliphate with its emphasis on science, math, and medicine that paved the way for our modern world. These breakthroughs made modernity happen. There was also a unparalleled mathematical expression of beauty in form as seen in the great mosques in Iran and the Alhambra in Spain. As we move into the future, we can appreciate and make proper use of the gifts of the past, incorporate the best, and move forward.

Thesis, Antithesis and ***Synthesis*** takes on fresh meaning within this timeframe. Yes, history flows in cycles. What is suggested here is another rhythm, one conditioned by the transits of Venus. This additional cycle is seen in a syncopated rhythm of 105.5 years plus 8 years, and then again in 121.5 years. The entire set involves a field of activity within which events and the people within those events are challenged with the inflow of great light. Old forms are swept away. Light sweeps in and transforms all in its path . . . science, art, politics, economics, education, religion, and the entire social-cultural spectrum. In effect, the transits provide a display of a great cosmic drama in which the planet Venus plays the leading role.

Thesis, Antithesis and **Synthesis**—in fact, these three movements are one of the ways we can approach the three gunas, the basic energies of the cosmos. This is also the way in which consciousness opens from levels of sattva, tamas, and rajas, and gradually moves out into larger understandings of life. Once the general public understands this, it can be a big step forward in terms of establishing a new philosophy and engendering a fresh approach to brotherhood.

The final synthesis of light, love, and power will require the integration of the best from our past legacy—the part that is evocative of Venus and the mind, intellect, and lucidity of explanation. Additionally it will include the new, fresh, and astounding ideas and discoveries that Venus promises to bestow.

Looking to the future, we are aware that in an even larger and greater cycle, the precession of the equinox, the 21st century is a transitional period. There is no doubt that a transforming civilization will use the energy of Aquarius, through the spirit of brotherhood, the light of illumination from Taurus (Venus) and evolve in the already existing field of manifestation, Pisces.

The above model of a Transforming Civilization is interactive and interoperative on the individual, collective, and global scale. Some of the areas in the United States to be affected in the transformational shift include:

- Renewed effort on the past of the United States in the area of Cultural Diplomacy.
- Revitalizing American volunteerism abroad in the spirit of goodwill.
- U.S.A. and China (along with Japan) commit to a leading role in Non-proliferation.
- Re-democratizing the political environment.
- Resurgence of the "dirigible" for long distance air travel and for transport of cargo.

- Analyzing, treating, and transforming Dr. Martin Luther King's "giant triplet" of racism, extreme materialism, and militarism.
- Expansion/Invigoration of innovative research and development
- Magical scientific breakthroughs in energy sources and usage.
- Transformation of the financial sector
- Emergence of Co-operacy

The term and phenomena of Co-operacy suggests the reinvigoration of social and political relationships. Venus energy will be in play and so it is useful to look at the Fifth Ray Method, that which is associated with Venus, since it will be employed.

> Let the three forms of energy electric pass upward to the Place of power. Let the forces of the head and heart and all the nether aspects blend. Then let the Soul look out upon the inner world of light divine. Let the word triumphant go forth: "I mastered energy for I am energy itself. The Master and the mastered are but One."
>
> Key to the Fifth Ray Method

Cooperacy suggests an integration and utilization of the light of Venus using the capacity to think, to discriminate truth from error, to employ technical expertise, to define issues clearly and to create new and better thoughtforms, and the ability to think and act scientifically. We know that energy follows thought and that energy follows consciousness. With greater awareness, the overshadowing solutions and prevailing conditions can fuse and unify. Consciousness will find what it needs.

All areas of life will be impacted but first and foremost education. This inflow of Venusian capacity will uplift all related streams of activity which are associated with an educated civilization. Education will be seen as the driving force for social transformation and in this sense the Transits of Venus pave the way for The New Age, The Age of Aquarius. Current grassroots protests represent early attempts to highlight changes. As these demands refine, and Venus will create refinement, they will lead to significant transformations here and abroad.

Rays from Venus will flow down on all of these as if in a constellation.

Venus
Education
Business
Government
Environment
Culture of Peace
Art and Music
Healthcare
Military
Media
Wealth of the Nation
Cooperacy

Table 10. The Flow of Rays from Venus

And so we approach the second transit of Venus in the face of many dire concerns—much has been predicted in terms of the End Times. There is the possible encounter with a comet or asteroid that could pulverize the earth; there are the Biblical prophecies. There is the buzz that solar flares may create genetic changes; there are rumors of crustal displacements or irresistible planetary movement toward a strange attractor. But there is an even more amazing prediction.

Certainly, the most revolutionary of all the predictions concerns that of an acceleration toward a great leap in consciousness for humanity as the beauty of Venus draws us to her. If this shift is to happen, Venus will have fulfilled her part in the creative process of manifesting a new civilization—one that identifies with the good of the Whole. With the will to good operating, world sorrow can diminish. The fruits and benefits of Venus will not be ignored; there will be a major reordering of principles in sequence: Unity—Peace—Plenty.

Such a civilization will incorporate and use as foundational stones the New Age Laws and Principles:

The Law of Right Human Relations
The Principle of Goodwill
The Law of Group Endeavor
The Principle of Unanimity
The Law of Spiritual Approach
The Principle of Essential Divinity

Table 11. New Age Laws and Principles

Figure 38. Planetary Lotus

The petals of the planetary lotus open as we cooperate, support, and share. The right design emerges and the Great Ones set a seal of approval and acceptance upon it. Humanity finds a way to quickly respond to the overall meta-paradigm, and the Age of Aquarius can begin. We are not there yet, but we must assume the role of pathfinders.

This story, the Transits of Venus, has employed not only history, but astrological symbols in an attempt to demonstrate the cosmic process of earthly existence. An attempt has been made to suggest to the reader that the entire cosmos is a teacher, a cosmic mind that can lend answers and provide energetic patterns of assistance.

There is a real possibility for the human traveler to achieve planetary consciousness, cosmic consciousness—one that will be enhanced and amplified by the Venus Transits. It is a story of love, light and power. And it is one of freedom.

Man/Woman is not conditioned by the stars. Energies are available. One can say in the spirit of Dane Rudhyar, that events do not happen to a human being . . . he/she happens to them.

The following themes emerge out of the incoming pattern:

Agriculture

Farmers worldwide have been salvaging seeds to maintain a viable source for planting in the future. Seed banks will be created and maintained and this will prove to be one way to avoid the risk for crop failures due to climate change. There will be a renaissance and expansion of small-scale localized agriculture. This has already begun and will escalate in the coming decades. Lessons have been gleaned about industrial agriculture that depletes soils, disempowers and displaces the people of the land. Within this milieu, ancient and indigenous methods will regain prominence and offer solutions.

These lessons are cumulative and global, often based on the Ancient wisdom of native peoples. Programs based on education for future directions will revitalize afflicted areas and restore life. Marta Benavides, Woman Peace Maker from El Salvador does not equivocate. She states, ***"We must be clear as to WHAT WORLD WE WANT and DECIDE TO LIVE FOR IT."***

The Moon

The moon will become the Persian Gulf of the 21st century. Moon bases will be used for initial exploration. Geology studies will pursue helium3 which will become a source of power. Research of this kind will provide incentives to develop nuclear fusion.

The moon will be mined not only for helium3 but also other useful minerals.

Tourism on the moon will develop and offer humanity a chance to view the home planet from a fresh perspective. Geodesic domes will be employed to provided a comfortable setting for such viewing.

Energy

Nuclear Fusion will provide more than enough energy for all of earth's needs. The terrestrial energy industry will be born with moon based plants to mine and process helium 3.
Nuclear plants using fission will continue in the interim
Renewable energy will expand and grow in usage and importance.

Rivers Streams and estuaries

Around the world, there are already regional meetings to address disastrous floods and the rising tides. The inflow of water has threatened settlements not only in Central and South America but also in India and Pakistan (the glaciers in the high Himalayans are melting.) Normally locations along the coast and in lands flanked by sea and estuaries have been ideal places to live with abundant fresh water as well as ocean seafood available. But with unprecedented rainfall events, these communities will make plans for mitigation and adaptation. Innovation and transformation are called for but there will also be some migrations of people.

Floating cities

There will be a renewed interest in Buckminster Fuller's concept of "Floating Cities. Australia would be a good place to experiment due to the largely unused and hostile interior of the continent. Australia has certain additional advantages for this kind of experimentation. It has a friendly immigration policy, has dealt well with its debt and has a 4% growth rate. A solid economic basis of this kind will nurture the kind of scientific advances that will and must emerge. Australia will experience success in many breakthrough advances and demonstrate a variety of new forms of living and learning.

Opportunities as well as crisis will arise due to global warming and rising sea levels. The people from the Maldives, for example, will someday loose their island kingdom and they may be ready to take the risk and move to the sky. Such experiments in living will put the principle of tensegrity to use—big strong flexible cables that will connect the floating cities in the air to the earth. Such cities will be environmentally controlled by the use of geodesic domes.

Fuller demonstrated the practicality of the geodesic dome in the 20th century. Now, with shrinking land there will be a renewed interest in his concepts and his encompassing ideas of restructuring human living. For example, in a business organization (micro) or on the Space ship Earth (macro) many processes are interrelated with a myriad of input and outputs. Local sources of trouble—civil war and famine—can and must be eliminated, thereby allowing life in those regions to become stable. Once stable then innovative, new methods can lead to improvement. We will profit from the experience gained in the past, improve it and move toward a new attainment. In this vein, borders between nations and all that those frontiers entail will be re-thought as humanity gains a better hold on unanimity.

Cloud Nines

In addition to the introduction of Floating Cities, smaller versions called Could Nines will be developed. These are smaller versions of Floating Cities and will be one-half mile in dimension. They can be used as schools, laboratories and factories.

Other Possibilities

Other possibilities for the new era include flying dwelling machines, submarine islands and ultra-high-energy world electric grid.

A new kind of economics will drive the transformation process with an emphasis on a win-win paradigm for government, industry and education. The transformation will release the power of intrinsic motivation and thereby turn the course of industry, at least in the West, sharply upward.

Taken to the macro level, we will consider the universe as a self-regenerative process—its actions and interactions are governed by a complex code of generalized principles which we call the laws of nature. Here on planet earth, the success of humanity can be accomplished by taking advantage of those regenerative process which Fuller one described as "synergetic recircuitry" This is not to say we will master nature but rather come into a local harmonic convergence with cosmic time-energy behavior laws. Mental activity stimulated by Venus energy will bring about the intellectual acuity for these kinds of advancements. Influences felt from the Solar angels will bring corresponding spiritual advancements on the inner plane.

Within this scene, what is needed on a planetary level is a system of continual improvement. A system includes methods, strategies and plans. Without a system, the much needed fundamental changes in global management will not occur. And so it is that the emerging Venus pattern suggests that those concepts and methods can get the intellectual and scientific boost to develop and come into being. Vulcan will bring the tools.

These discoveries and innovations will fill the gap between what has transpired between the 19th century and the 21st century transits. What is possible and what is hinted at, is a new era in human history. It will involve a continually transforming system on planet earth with an emphasis on peacebuilding instead of war; conflict resolution instead of local skirmishes; prevention and education instead of incarceration; cooperation with the laws of nature instead of forcing control. Will we succeed? There is vast room for hope with this last transit but there is still the question of will.

Stories of 21st century women from all parts of the world and all religious and social backgrounds demonstrate how they have taken initiative at the grassroots level. Through their experiences, lessons are learned not only about the roots and components of violence, but also about positive self-generated actions that transform communities and eventually nations. It has been clearly demonstrated that women in these examples have experienced great hardship due to the harsh, violent prevailing conditions. They work in a variety of ways to ameliorate existing conditions and employ practical strategies. Their work has had local and international effect; they have influenced UN resolutions. For example, UN Resolution1325 addressed the disproportionate and unique impact of armed conflict on women who are, along with children, 90% of the victims of war. It also addressed the importance of women's equal and full participation as active agents of peace and security. UN resolution 1820 addressed the issue of war and sexual violence. It is clear that women are on the map of human consciousness and this will only increase.

Themes such as these that have been introduced during the last eight year phase, 2004-2012, will be played out in the next 105 years. Humanity will find ways to self-organize and renew itself drawing from its innate integrity and boundless energy. With the appropriate use of the almost unlimited power that will be available through nature and technology, mankind can, within a generation or two achieve a high standard of living for all inhabitants of the planet. International cooperation is key.

This last transit, in 2012, will carry us forward up to and into the New Age. Venusian energy will prepare us to become fit to perform the task of the future evolution as we approach the Age of Aquarius. Within this mix of events, synthesis is key. We will not reject the past but build on it, transform it, and cooperatively create the Age of Aquarius.

> And whether we are to be
> A complete success or utter failure
> Is in such critical balance
> That every smallest human test of integrity,
> Every smaller moment-to moment decision
> Tips the scales affirmatively or negatively
> Of Total human success
>
> Buckminster Fuller.

Fuller's last book, **Critical Path**, traces the origins and evolution of mankind's' social political and economic systems to show how we got to the present situation. He goes on to describe how we can meet the ultimate challenge to our future. He describes himself as a comprehensionist and warns against the current emphasis on specialization. His research indicated that extinction is a consequence of over-specialization, for the latter bred out general adaptability. Humans, because of the gift of mind, have access to general principles in order to support the integrity of an everregenerative universe. One can with little imagination transpose this kind of thinking to what happened in Atlantis and lessons can be learned.

It becomes clear that what is incoming is a new phase in our history. It will reflect a balance between ego-system and ecosystems. Will we do it? Will we create a new era in human history?

The next transit of Venus will be in 2117.

At that time a new set of themes will be introduced and firmly set in place. Many deep and structural changes will have occurred by 2125 and humanity will work those out up to and until the next set of transits. As in all cases discussed in this text, the transits of Venus can be seen as book-ends in time when themes are introduced and played out in a specific time frame. The 22nd century transits of Venus will mark the beginning of the Age of Aquarius and will therefore be a major step forward in terms of human development. But we should pay much attention to the present transits, as they are the prelude to the New Age.

Things We Can Do

- We can use the energy from the current transits to be open to intellectual activity, to be aware of beauty and to celebrate the tempo of the cosmos that never ceases to surprise and to challenge us to be more than we are.
- We can be open to the new and fresh approaches which will develop.
- We can support and participate in the elevation of women to positions of influence, power and respect.
- We can consider our planet as a space ship earth with certain limitations and requirements for health.
- We can celebrate the unanimity and basic rights of all people.
- We can also emphasize the need to use the mind to guide human affairs to control human and earthly affairs.
- We can transcend separatism and fear.

These initiatives will move us forward in a way that will support human-advantage increases as we encounter the inexorable emergencies that lay ahead. At the same time that they will nurture the health of the planet itself.

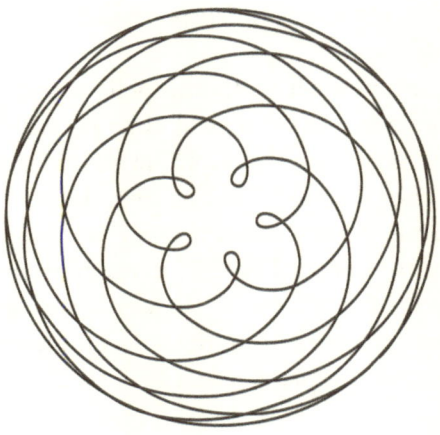

Figure 39. Movement of Planet Venus

Magical geometry is seen from the geo-centric perspective as Venus weaves a pattern of a five petalled rose. It has been said that every eight years Venus and Earth kiss each other and a new petal is formed. The Cosmic dance with retrograde motions and intricate cycles present a symbol that resonates with the very essence of Venus. The rose reveals her as the goddess of love, beauty and clear thinking to humanity.

About the Author

Gail Dimitroff holds a PhD in Leadership and Human Behavior. In 1980 she chaired the County Commission on the Status of Women and is an activist for women and women' issues. She has taught courses in Women in Management and Stereotypic Images of Women in Media. Gail studied short story writing with Paul D. Peery who for many years held a writer's workshop in Coronado, California. She has won awards for short story and article writing with the San Diego Writer's and Editor's Guild. Her work with Mr. Peery and her interest in the short story form paved the way for the writing of her Memoir, Mande, which falls into the genre of literary memoir. She has lectured extensively on the Memoir form and has presented at Border's etc. She is a member of the National Penn Women's Association.

As a business consultant in the field of continual process improvement Gail taught, published and lectured in North and South America. Gail considers herself to be a spiritual peacemaker and holds group meetings at her home in La Mesa, CA, at the time of the Full Moon for world peace. She is an active participant in AAUW, VOW, and a graduate of LEAD San Diego. She was one of the first students at the San Diego College for Women, now the University of San Diego. Gail was named the 2011 philanthropist of the year by the San Diego Women's Foundation.

Gail's interest in Cosmology, Astrology and Esotericism has

peaked since she became a faculty member of Morya College—an on-line school where students study the Ancient Wisdom. There she teaches Esoteric Psychology and Astrology as well as a study of the Deva Kingdom. At her home, she teaches a class in Esoteric Astrology and Soul Dynamics.

Bibliography

Works about the Ancient Wisdom Teachings by Alice A. Bailey
There are at least two dozen books available from Lucis Trust Publishing Co. P.O. Box 722, Cooper Station New York, N.Y. I recommend the following in terms of this present work.
Esoteric Astrology, 1951.
A Treatise on White Magic, 1951.
The Rays and Initiations, 1960

Other Works about the Ancient Teachings

Blavatsky, H.P. *The Secret Doctrine,* vol. 1-3 Los Angeles: Theosophy Co., 1928
Agni Yoga, 1929, Agni Yoga Society, Inc. New York, N.Y.
Heart, 1932, Agni Yoga Society, Inc. New York, N.Y.
Thoughtline. November 2003, Arcana Workshops, Culver City, CA
Triangles Bulletin, No. 146, December, 2003, Lucis Trust, New York, N.Y.

Other Works about the Seven Rays

Burmester, H., *The Seven Rays Made Visual,* Marina Del Ray, CA. 1986

Other Works about Astrology

Data for the Sagittarius rising U.S.A chart was derived from khaldea. com

Data for the Scorpio rising U.S.A chart was derived from homestead. com

All other factual data regarding charts had been derived form Lois Rodden's <u>ASTRODATABANK</u> 25 Raymond Street Manchester, MA 01944-1614

All charts were constructed using Solar Fire Program.

Carter, C. E. O., *The Astrological Aspects*, L.N. Fowler & CO. LTD. London England. 1930.
Oken, A. Alan, *Soul-Centered Astrology*, The Crossing Press, Freedom Ca. 1990.
Rudhyar, D. *The Astrological Mandala, The Cycle of Transformations and Its 360 Symbolic Phases*, Random House, New York. 1973.
Saraydarian, Torkom., The Psyche and Psychism, T.S.G. Publishing Foundation, Inc., Cave Creek, Arizona.

Other Works about the Transits of Venus

Hirshfeld, Alan W., *Parrallax, The Race to Measure the Cosmos*, W.H. Freeman and Company, New York. 2001
Maor, Eli., *June 8, 2004, Venus in Transit*, Princeton University Press, Princeton, New Jersey, 2000.

Historical Facts and Dates

The Columbia Encyclopedia, Fifth Edition, Columbia University Press, Houghton Mifflin Company, 1993.
The Great Books of the Western World, William Benton, Publisher, Robert Maynard Hutchens, Editor in Chief, Mortimer Adler, Associate Editor
Europe and Western Civilization in the Modern Age, Professor Thomas Childers. The Great Courses Series. The Teaching Company, Chantilly VA. 1998.

All quotes taken from "Quotables" on the internet.

Appendix I

Transits of Venus: 1601-2200

Date	Universal Time	Separation
1631 Dec 07	05:19	939 "
1639 Dec 04	18:26	524 "
1761 Jun 06	05:19	570 "
1769 Jun 03	22:25	609 "
1874 Dec 09	04:07	830 "
1882 Dec 06	17:06	637 "
2004 Jun 08	08:20	627 "
2012 Jun 06	01:28	553 "
2117 Dec 11	02:48	724 "
2125 Dec 08	16:01	733 "

Fred Espenak web site, NASA

Appendix II

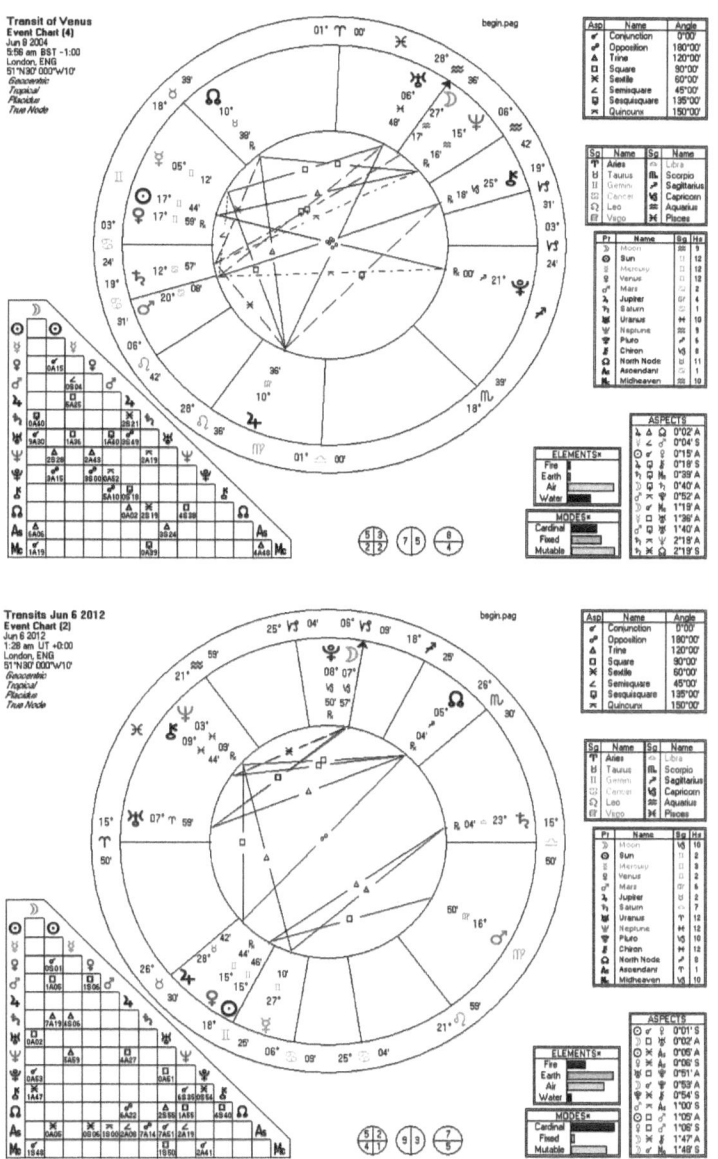

163

Appendix III

There are frequently disputes regarding the charts of Nations and it is no different for the chart of the United States. Numerous birth-charts for the United States have been given and despite innumerable rectifications there are still conflicting opinions. This is in no way an argument for any one of chart. Three charts for the United States have been given so that readers can use the one that draws them. The first is the most popular chart with a Gemini Ascendant.

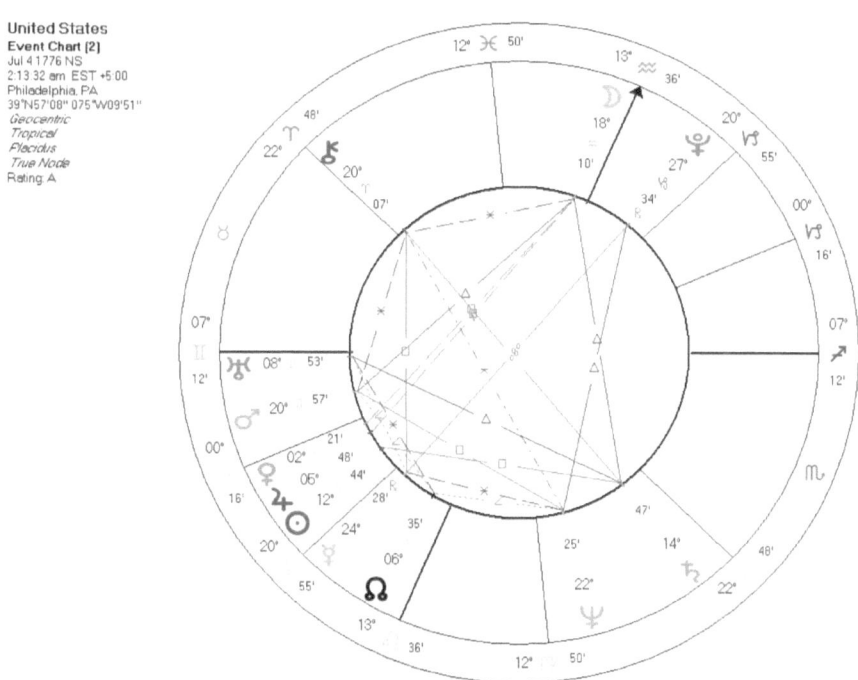

United States
Event Chart (2)
Jul 4 1776 NS
2:13:32 am EST +5:00
Philadelphia, PA
39°N57'08" 075°W09'51"
Geocentric
Tropical
Placidus
True Node
Rating: A

This chart resonates with those who see the Gemini Ascendant appropriate to the youthful, mobile and curious people of the country. Add to that the various innovations in many fields including communication and a good case can be made for this chart where Uranus conjuncts the Ascendant. Freedom is associated with Uranus while Mars first house place men indicate self-rule. The stellium indcates the enormous wealth of the United States and it is this chart that has been used in the discussion regarding the transits of Venus in this book. The transit have been marked on each chart.

The second most popular chart of the United States is that with the Sagittarius Ascendant. Dane Rudhyar made a strong case for this chart. One can cite the Americans' love of freedom as well as their outgoing, generally friendly nature congruent with the Sagitarius Ascendant.

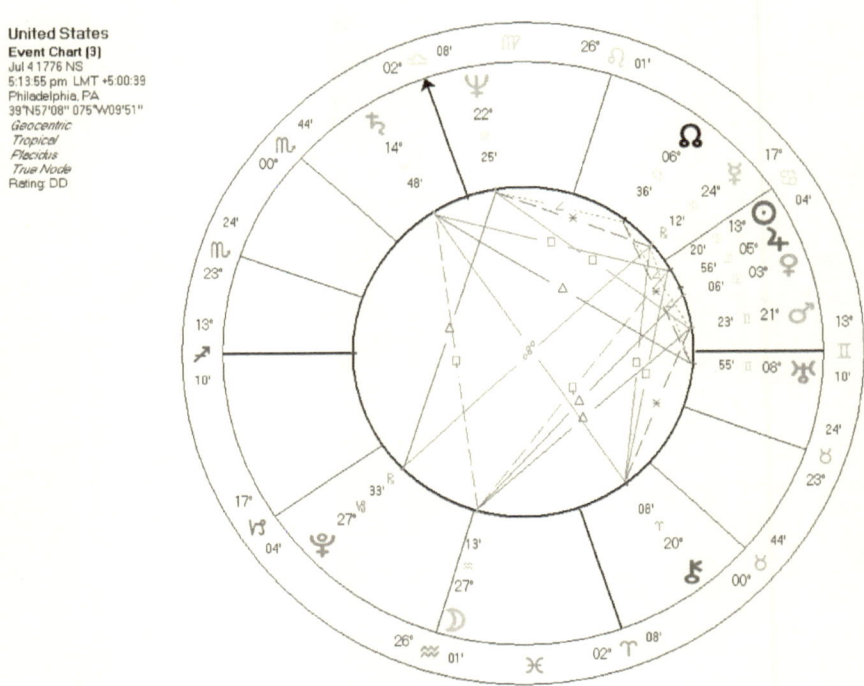

United States
Event Chart [3]
Jul 4 1776 NS
5:13:55 pm LMT +5:00:39
Philadelphia, PA
39°N57'08" 075°W09'51"
Geocentric
Tropical
Placidus
True Node
Rating: DD

The third chart given is that with the Scorpio Ascendant. Scorpio defines the American personality as terms of power, self control, interested in financial powers. The capitalist system is often equated with the United States. Furthermore the nation's symbol is the bald eagle and the eagle is one of Scorpio's symbols.

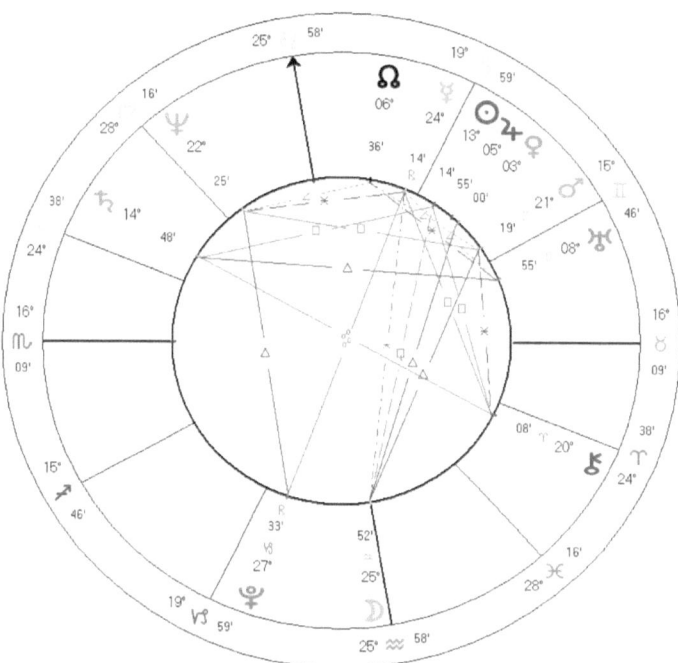

USA
Event Chart (2)
Jul 4 1776 NS
4:00 pm AST +4:00
Philadelphia, PA
39°N57'08" 075°W09'51"
Geocentric
Tropical
Placidus
True Node

167

INDEX

Y

Z

www.ingramcontent.com/pod-product-compliance
Lightning Source LLC
Chambersburg PA
CBHW032005170526
45157CB00002B/550